HONEY FARM DREAMING

A MEMOIR ABOUT SUSTAINABILITY, SMALL FARMING AND THE NOT-SO SIMPLE LIFE

ANNA FEATHERSTONE

CAPEABLE PUBLISHING

Published by CapeAble Publishing

First published in Australia in 2018

To order directly, please order online at www.honeyfarmdreaming.com

ISBN: 978-0-9807475-4-6

Cover design: Gayna Murphy, Mubu Design

Cover image: Hung Quach

Interior: Vellum

Printed by: Ingram Spark

A catalogue record for this book is available from the National Library of Australia

To Andrew, you carry ideas in your head, our kids on your shoulders and always – the kindest of dreams in your heart.

There are moments when the heart is generous, and then it knows that for better or worse our lives are woven together here, one with one another and with the place and all the living things.

— WENDELL BERRY, JAYBER CROW

CONTENTS

INTRODUCTION

Our farm life began in 2006, the year terrorists had humble toiletries banned from aircraft, Cyclone Larry peeled the Queensland banana crop, and Norway began building the Svalbard Global Seed Vault in case of agricultural Armageddon, which, unfortunately, has since sprung a leak due to a melting glacier. With husband Andrew recovering from Sarcoidosis, a disease that confounds even the medico's on Grey's Anatomy, we hired a giant bug-out bag of a removal truck, packed three kids under six and a crate of stuffed animals in the car and set about creating a life on the land. I was thirty-seven, Andrew forty-one, and we were getting out of Dodge, hunting the simple life, ready to raise the one, three and five-year-old free range.

We were ready to grow and smell the roses.

Years earlier we'd been spurred into thought – though not action – by 9/11. It was like Bin Laden bombed our city blinkers off. We began researching the Middle East, oil wars, and how dependent the West has become on black gold to farm and ship foods vast distances. One smear led to another and we discovered agricultural giants were spraying the life out of life itself, and that humanity was mining its own future.

Meanwhile, our own lives were leaking away in city boardrooms, our combined smarts solving not much more than increasing the gap between rich and poor.

We couldn't afford to live like that anymore, and though we couldn't afford the verdant hills of the Southern Highlands, or the millionaire celebrity pastures of Byron Bay, we could afford eighty-nine acres outside Nabiac, a highway-bypassed village thirty minutes to anywhere, earthed at the southern end of the fertile Manning Valley and on the western border of the sandy Great Lakes – a place that Death Adders also like to call home.

Within weeks of buying the farm, bills started spawning in our mailbox. Fencing steel bolted twenty-five percent thanks to voracious Chinese demand. Drought, bushfire and diesel prices stacked fifty percent onto lucerne hay. Privatisation and profiteering vaulted electricity prices fifteen percent. The old tractor and quadbike, bought with the place, sputtered for repairs, the hot water system went cold, a mouse took out the oven electrics and we blew our first month's budget by eight hundred per cent.

We quickly discovered the region was practically barren of food, farmers and career opportunities for people with our skill set. Except for the third-generation dairy down the side road, many landholders nearby worked jobs up to an hour away, some subsisted off superannuation, pensions and welfare schemes, and there were rumours of a couple of insurance scams. Others ran steers to keep taxes and the grass down, and out in the hills there were regular raids on farmers of the illegal leaf variety. Four of the shops in the tiny village were closed; paint and optimism peeling.

It hadn't always been the case. According to Nabiac historical records, back in the 1870s steam powered, paddle wheelers with names like *Dauntless*, *Nil Desperandum* and *Wheelbarrow* (you even needed a sense of humour back then!) plied the nearby Wallamba River taking the region's timber, possum skins and kegs of salted butter via the Pacific Ocean, to markets in Sydney. As late as 1969, a riverboat navigated the Wallamba collecting milk from the thriving dairies along its banks. Then the diesel-powered road tankers took over, the industry deregulated, the price of milk evaporated, and the bulk of the small family dairies went down the drain.

Our farm's previous owner of twenty years, a Mormon minister, installed whirlybirds, skylights and redemption for a living – but one day he saw even extra light, building two weatherboard cottages on the property to rent out as cheap farmstay accommodation.

That's how we came to farm people, and the depths of our resolve. Our main crop never ended up being beetroot or honey, turmeric or rosellas, miniature Galloways or goats, Dorper sheep or Sussex chickens – though we produced fields of them every year, selling at farmers' markets from Sydney to Port Macquarie. Our main crop became the biggest, baddest, beast of them all: the voracious, unpredictable, and often times astounding *homo-sapien tourista* and its sub-species *backpacker bewareus*.

This is just some of what happened.

WANNABEE

I'm allergic to bees but I want to be a beekeeper. It's a bit like being morbidly obese and applying to be a mystery shopper for McDonalds, or being a British albino and taking a job as a tanning bed tester. It's not really wise, it's not really something a mum of three should do, it's not really meant to end well. It's not really on the scale of snorting coke in the office bathroom or sleeping with strangers or base jumping off the Eiffel Tower or having a nut allergy and ordering a satay, but it is on the scale. Maybe beekeeping is just a green-inclined, thoroughly respectable way of walking on the wild side.

I remember the first and last time I was stung by a bee, I was thirteen and sandalling it down to the beach. There was no footpath, just a lawny, dotted-with-clover park leading past the lagoon to the ocean. I never made it into the surf that day, the clover-loving bee met sandal, met toe, met squish, and I met the radiating pain of an intensely burning fry pan embedded in my skin. My entire foot blew up to the calf, wept fire for a week and I couldn't fit a shoe back on for a month.

That's why I went to the doctor's today, to get the final okay.

"What did he say?" asks Andrew.

1

"I should be okay if I don't get stung near my throat. And he says to keep an Epi-pen on me at all times."

"You sure you want to do this?"

For months I'd been reading about honeybees dying across the planet from the Dracula-mimicking insect Varroa Mite, and from toxic agricultural chemicals pumped out by the world's biggest corporations. I'd drunk in information about the fake honey flooding the US, the synthetics out of China, the corn syrup from Turkey and the plight of Australian-bred Italian honey bees handed death sentences via work visas and one way QANTAS tickets to the mega-monoculture almond groves of California.

"If I have bees I can help save them. Plus, if we want resilience on this farm, we need pollination. We want to feed our family? We have to have bees!" I say to Andrew. "Plus didn't some smart guy, Einstein or someone, say 'when the bees die, humanity only has three years to live'?"

He looks at me calculating what it's going to mean for his already heavy workload if I get stung, die and leave him the apiary responsibilities too.

"Don't worry, you milk the goats and I'll milk the bees. Strict separation of responsibilities, that way neither of us has to do everything. And I promise to protect my throat," I gasp, pretending to strangle myself.

In the past few weeks, I've devoured four books on honey bees, aka Apis Mellifera, and sat up straight in a two-day beekeeping course held by the Department of Primary Industries. Here I'd learned less about killer bees and more about everyone who wants to kill them: Tropilaelaps mite, Tracheal mite, Braula fly, Small hive beetle, Chalkbrood disease, Varroa mite, Nosema – the dysentery of the bee world – and one of the deadliest assassins of them all, American Foulbrood Disease.

This disease, spread by century-surviving killer spores of Paenibacillus larvae, likes to turn bee larvae into snotty, stringy corpses.

It's baby on baby UFC.

Bacterial cage fighting with only one possible winner.

In Australia, AFB in your hive means you're legally required to murder your bees before they spread the disease to other hives in the area. Apparently, a bottle of petrol into a hive at night can fume fifty thousand dead in minutes. Then you either have to bonfire the hive in a pit then bury it, or send it, wrapped in plastic, to a nuclear facility in Queensland to be gamma irradiated. I nearly threw down my beekeeping gloves at the thought. But then the lecturer moved on to the healing power of honey, the intricate 'waggle' dance bees do to tell their friends where the best honey is, and how educated beekeepers are really important for bees given all the pests and disease they're bombarded with.

Even after the course I was still too nervous to get bees, I didn't feel I could do them justice, I didn't feel I had enough experience to care for fifty-thousand lives. I didn't want their blood, or tiny little decapitated legs, or petrol sniffing lungs on my hands. I didn't want to be stung. I didn't want my throat to bloat. I didn't want to fail them, or the family. Then one day I saw an ad in the local paper, "Coastal Gold Honey: Honey, Swarms, Honeycomb – Call Rod the Bee Man." So, I did, but not for any of that; I called for expert hand-holding, and the hand-holder is pulling in now.

"He's here, the bee man's here!" I say, jumping up and down.

Rod's blue ute idles roughly in the driveway and I run out to meet him. His face shows him to be in his sixties, but when he gets out of the car to shake my hand he's got the body of a man in his thirties. He's tanned, with beekeeper-buff biceps, sculpted quads and blue eyes crackling with life. His face is open, but a bit pointed and I think he kind of looks like a bee.

"You must be the wannabee," he says warmly.

"That's me! Thanks for coming!"

"I've got your girlz on board." He even sounds like a bee!

I look to the back of the ute and see two white hives strapped together, forty bees buzz about having escaped the cardboard barriers wedged in the entries. "Where do you want them?"

"Under the big gum out on the hill," I point to the old Eucalypt five hundred metres away. It's a site to provide shade in the summer peak, nectar in spring, protection from the ripping winds of winter and a home far enough away they don't sting their way into the comings and goings of the house and farmstay.

"Grey Ironbark, best nectar, they'll like that. Good choice," he agrees and I go pink with I-got-something-right pleasure.

"I'll go get my suit on," I say.

"You don't really need it love, the girlz are in a good mood and you've already got jeans and a long sleeve on. Plus, it's toasty today."

"Well, I'm kind of allergic so I better."

"We're all allergic love, allergic to pain," he laughs. "See you out there."

I run inside and unfold my birthday gifted suit for the first time. I thrust my jeaned limbs through the elasticised ankles, my clad arms through the tight wrist bands and carefully zip the suit to my neck. Then, like I'm veiling up to get married to Matt Damon in The Martian, I pull the mesh hood over my second-line-of-defence cotton hat, pull the two zips of the hood around and Velcro them together so there are no gaps for bees to make their way to flesh. I tuck my thick-as-two-bricks socks over the pant legs, then slip into my knee-high gumboots, drawing protective white gloves over the sleeves to my elbows. And, like a Teletubbie walking on the moon, like a nuclear scientist picking through Fukushima, I head out across the paddock to Rod, who, in shorts, thongs and light cotton singlet, wipes sweat from his forehead and says, "don't worry love, I'll call an ambulance when you pass out."

I can feel my back sweating itself to my shirt, my feet going slimy in my socks, but I feel safe and calm in my bee-repelling get-up.

"This here is the brood box," he says, lifting the first white rectangular box from the ute. "I've put in four frames of eggs, four frames of honey and twelve handfuls of worker bees. The queens arrive?"

I shake my head.

"Bugger! They'll get here today then and when they do, give 'em water and keep 'em warm." He gestures for me to unblock the entries.

I bend over and my hat slips down under my hood blocking sight from my right eye. My fingers flail inside my gloves so I can't get a grip. I feel like a visually challenged, inflatable sumo suit wearer trying to thread a needle.

"You make things hard for yourself when you wear all that," he says, "and for the bees."

"I know, but I don't want to get stung, not just for me, but I don't want to kill them."

"I hear you," he says, "but when you work bees, fewer die if you work fast and sure."

I fumble around faster, finally removing the little door blocks.

Little legs, little bodies begin to appear. It's like I've lifted a roller door on another world. The trickle of insects turns into a yellow and black waterfall and they spill out, the buzzing and humming vibrating my eardrums like a sparkly electric wave. Out of my left eye I see some bees taking straight to the air like Sikorsky's, others rise slowly like hot air balloons, while others use their segmented legs to SWAT-team it to every outside corner of the hive. A few launch at me and I take a step back.

"Get closer, get in there," urges Rod, "take a good look through that hood of yours, they're yours now...you're theirs."

I shuffle forward to the ripples, the pulsing sound enveloping me, then I focus my left eye. I'm shocked. Within the ripples are individuals...furry, soft, friendly, sharp, carved, top models, country cousins, sprightly, sage, sitters and heavy hitters.

"They're so different looking," I say.

"Yup," he agrees good naturedly, "like Asians."

He reaches his hand down and lets a bee walk on his finger. "Treat 'em good and they'll treat you good." He brings his finger to my glove and the bee walks onto my hand, I'm mesmerised by her steady calm, her un-sting-ingness.

"Uh oh, I've seen that look before," says Rod, "you're not going to be stopping at two hives love. I started with two and now I'm at two fifty."

"Two hives are plenty, I just want to keep them alive. And me."

"We'll see, but your main problem, besides your hat," he says pushing my cap back into place and making sure my hood is straight, "is going to be that bloody African hive beetle. You gotta get out here every week and squash every damn bugger you can. If you don't they'll shit everywhere and your girlz will leave. Got it?"

I crouch down slowly, making sure there are no bees to crush in the folds behind my knees. I feel unsteady and it's not the heat.

"Do you think you could come back tomorrow Rod? I don't know if I can put the queens in on my own."

He looks at me slowly, taking in the gumboots, the full rig, my zipped-up self.

"Sure, we'll make you into a beekeeper, not one of those bloody bee-havvers. Get their hives, don't know a bloody thing, then there's bloody beetles, angry bees, dead bees, foulbrood. Should be a law!" He pauses. "Sorry for raising my voice ladies," he addresses the bees. "Jump on love and I'll give you a ride back, there's nothing more we can do today and I gotta get going."

I sit on his tailgate enjoying the breeze hitting the sweat, enjoying knowing I won't be on my own tomorrow. At the gate, I slide off to let him through and he sticks his head out the window.

"And get that bloody suit washed or it'll go mouldy with all your juice."

"Got it boss," I say smiling.

At the house, I peel off all the wet layers and throw them, hat and all, into the wash. I pull on another pair of jeans, a shirt and drive straight to the village. The post office is a small counter inside the Old Bank Centre, and we've discovered it's Nabiac's one-stop shop for everything from lettuce to lottery tickets, bread to beer, mouse traps to magazines. I stop briefly at the outside community noticeboard, layered in handwritten notices, to see if there are any second-hand kid-sized beekeeping suits for sale:

LOST: after the storm, much loved cattle dog cross, answers to 'Dog'.

FOUND: after the storm, Miniature stallion. Whoever owns him owes us a bale of hay, he's a guts.

MISSING: from your life, Nabiac Library. Reward: 15 loans of books and DVDs.

WANTED: tick-eating guinea fowl x 3.

FOR SALE: home brew set up plus 10 000 bottle caps.

FREE TO GOOD HOME: 8 guinea pigs, related.

I head in the screen door, walk past the magazine racks and to the Australia Post counter. Colleen the post mistress looks at me nonchalantly.

"I'm wondering if any mail has arrived for us?"

"It has," she says turning to the racks behind, "and it's buzzing." She gently hands over an express post satchel punctured with breathing holes.

"They're bees," I say excitedly.

"I figured," she smiles.

"Queens!"

She blinks her eyes. Once. Twice. Another smile flits across her face. It's polite post mistress Morse code for, "I know you're new to the area and you're excited, but there's now a line of people behind you, it's time to move on."

I position the bee package in my hands, like the queen jewels on a crimson pillow, and slide a few steps to the right to the retail shop counter. It's lined with lottery tickets at one end, lollies at the other, and is looked after by Lesley.

"What can I do for you?" Lesley asks.

"I need a lighter," I say, then backtrack, not wanting her to get the wrong impression, "not for smoking, for my bee smoker!"

"Right."

"Queens!" I point to my package.

"Well, have fun with them...this," she says, passing over a BIC, a gaseous gadget that feels way too comfortably familiar, the weight in my hand transporting me back to my teenage years when everyone seemed to be lighting up.

I want to rush to the car but it's like I need to solemnly walk to give the bees respect. An old man opens the door for me.

"Queens!" I say.

"Be better with a republic," he mutters. But in my head, I'm hearing royal bugles, golden trumpets and the clip clop of bejazzled white horses leading the queen bee parade.

In the car, I gently tug at the adhesive until the package comes apart. Inside are rolls of cardboard held together by rubber bands. Bzzzzz. I slip them off and the cardboard unrolls revealing two timber boxes, each just a little longer than a matchbox. Bzzzzzz. There's flyscreen stapled on one side and through it I can see the pacing of four worker bees and the longer, taller, regally striding queen. Bzzzzzzzz.

"Hello! Welcome! My name's Anna. I'm going to take you home and get you comfortable. This is so exciting! I hope you are excited! I am honoured! Thank you for coming." Bzzzzzz Bzzzzz.

I gush at them all the way home then parade stiffly into the house pretending I have a book on my head, that I am the incarnation of a carpenter's level, determined not to tilt and upset the queen and her courtiers.

Give them a drink, that's what he said. How does one give bees in a box a drink? Through a fine straw? With an insect-sized eyedropper? A microscopic chalice? I flick through my trove of books and discover I just need to transfer two drops of water from my finger to the flyscreen. The droplets sit there like little bubbles and the worker bees extend their tiny tongues – proboscises – to sup.

Keep them warm, that's what he said. Make them tiny ermine cloaks? Crochet a rug for their multiple knees? Slide miniature Ugg boots onto each of their six legs?

I decide to take the two boxes to the bedroom and make them comfortable on my pillow, then pull the quilt up half way over the boxes so they can still breathe. There are bees in my bed and I am so outrageously happy and in wonder I get in with them. Bees and me in a bed. Insect kink. Bzzzz. Bzzz. Bzz. I lie still and they finally settle. Zzz.

The next day Rod returns. And then every week for four weeks he literally holds my hand as I suit up. He mentors me on the best way to light a smoker, the right amount of pine needles to use, the way to puff it through the front door so the bees chillax. He stands by in his King Gees while I lift the lids of the hives, gingerly lifting out frames, holding them to the sun to spot tiny eggs and to revel in the development of the fatty white pupae. He teases me about my protective get-up and reminds me every beekeeper gets stung... eventually.

I feel alive as I learn from him. It's the combination of his brimming love for bees and the oxygen that comes from listening rather than speaking, the thrill of learning compared to the responsibility of knowing. I wanted to be a beekeeper, but I'm also feeling the benefit of being kept, rapt.

He encourages me to be smooooth and aware as I open lids and lift boxes.

"You want to caretake them, not just take from them," he says.

Each week there are more and more bees, they surge in the hives like waves, and I talk to them, as Rod does, to keep all of us calm.

"Hello beautiful ladies, it's just me, coming to check on you. I planted some sunflowers for you today and some perennial basil and lavender. How are you going? I'm just going to put this clearer board on so you can get out of the honey box safely, watch your legs. Okay, I'll wait. Sorry to bother you but I'm going to give you another box today, Rod says I need to give you room because there's a honey flow on from the Ironbarks and you're going to want to sock even more honey away."

"I think you're speaking their language," says Rod that afternoon as I hang my suit in the shed to air. "Next time you're on your own kiddo."

"I don't think I can do it without you."

"You can't afford to keep doing it with me either," he says, acknowledging the bill I've been running up. "You're ready, you really are."

The next visit I do solo. I feel Rod's balmy presence even though he's tending his own bees in a forest twenty kilometres from here. The bees respond, docile as the dandelions nodding in the breeze.

"Thank you so much for this honey, I'm planting you more flowers," I say, gently brushing the few who haven't headed out through the clearer board from the honey-filled frames. As I seal the top lid and carry the treasure chest of honey to the ute, I feel like a grown up, not a risk taker; I feel responsible, not irresponsible; on a life-wish, not a death-wish.

That afternoon the whole family gathers in the shed for the harvest. I use a serrated knife to shear off the thin wax caps of the honey cells and the children suck the gooey sweet pieces as they fall. Then we put three frames at a time into the catches of the spinner, and crank, crank, crank till the whirring smell of summer, dew and blossoms rushes sweet and strong into our smitten faces. Shiny, golden, decadently bulbous honey balloons from the tank into waiting jars below.

My fingers still sticky, I phone Rod. "We did it! We harvested!"

"How'd it taste?"

"Like sunshine."

"That'd be the Ironbark. There's nothing like it."

"Rod, I've been thinking...about getting a couple more hives... just two."

"It was only a matter of time," he chortles, "I knew you'd get stung."

He was right, pierced right through by humble honey bees, pots of gathered pollen, petal rainbows, window panes of gold, and the sweetness of life.

* * *

HONEY ISN'T JUST GREAT TO EAT, IT'S ALSO GREAT ON YOUR FACE! FOR A honey face mask recipe so natural you can even lick it off, see page 217.

HANNIBAL LECTER JNR

luck. Crow. Cack-aw cack-aw. That right there is a chart topper. It's the sound of living the dream and it's the background beat at 6 am as I leap out of bed, raring to implement our mid-life-the-whole-world-is-going-to-pot-and-we're-going-to-stop-it crisis mitigation plan. Today we start making a living from ethical living and today we hope to make a difference.

There are eight hours to get ready for our first farmstay guests. Check in time is 2 pm and they've booked both cottages. Eight beds need hospital edges, eggs need collecting, udders need milking, honey needs bottling, lawns need taming, the pool de-greening and our own kids who are just one, three and five need parenting too.

Not a problem.

It's a dash. I zip about like Tinkerbell. Here one minute, there the next, sprinkling imaginary fairy dust joy, sprucing and sheeting, wiping and washing. Andrew follows behind straightening the sheets, reorganising the breakfast hampers, spotting the spots.

"How about you go mow babe," he says, demoting and releasing me at the same time.

It's a clunky beast, the old mower. Seat foam writhes out of the slashed upholstery, the metal skin is drizzled in grease like a fifty-year-old baking pan and you can chew on the cookies of black smoke sent up at ignition. But to sit on it, to drive it, is sheer bliss. Guilty pleasure because I'm using world-ending polluting diesel, but bliss because in addition to turning the chaos of snake-camouflaging grass into carpet, it's just so satisfying to see instant results, to visually and immediately achieve. You don't get that on the conveyor belt of corporate life.

I'm reigning from my belching throne and the vibes are off the scale positive. We're about to make our first income from the farm, our first income in months. Plus, we're going to meet people! Since arriving our only visitors have been two crusading flocks of Jehovah's Witnesses who were sent packing minus the tiny seeds of our souls.

Clunk.

Bugger!

My rule is over.

A geyser spurts twenty feet into the air from the water tap and pipe I've just crushed and slashed. Mains pressure.

Hope the guests aren't expecting showers.

"Andrew!!!?!"

What's with the whole Handy Andy thing anyway? Are all Andy's handy? Do they get lumped with the rhyme and have to skill up to it? Whatever, I'm stoked to have had this working model since we met on a work trip when I was 21.

We're both soaked by the time he tames the deluge. It's a flood Noah would have been proud of, but it's going to cost us, there are severe water restrictions in place due to the desiccating El Nino radiating in from the Pacific.

It's a little late for 'watch where you're going,' so he says, "maybe a bit slower next time."

And that's when we hear the gravel crunch.

Eight expensive rims roll down the locally quarried, blue metal driveway. Three hours before check-in, the guests have arrived.

Three boys rocket out of the cars looking like they're on advance security for Vladimir Putin. They're fast and low across open spaces, leap in and out of pens, edge around corners, test the boundaries. Finally, they circle back to the vehicles.

Still dripping, we head over too.

The tinted electric window of the first Beemer rolls down revealing a late-30's couple in blindingly white shirts.

"Welcome," says Andrew.

I head to the second Beemer.

"Hi there, I'm Anna. Ready for a fun holiday?"

The freshly blonded Mum and manscaped Dad thin-smile me.

"They're all yours, we'll go unpack," he says, and accelerates child-free toward the cottages.

I don't have time to mention I was just heading in to make our kids lunch. Or that I was about to change into dry clothes. Or that they are three hours early but I'm excited to meet back with them at 2 pm to give them their welcome tour. I don't have time to do anything before I hear a scream.

The self-sufficient, income-producing simple life has begun.

The scream shrills from the bunny pen. I try a casually nonchalant run in the direction of the miniature cashmere lop-eared furballs.

"Do it now! Spread it," commands the unusually cool voice of a six-year-old.

I speed up, fling open the gate and rush in on Hannibal Lecter Jnr stage managing his younger brother in a scene from a drug cartel torture movie.

The five-year-old has Snowy Rabbit stretched out on the ground like a mini-bear rug. Hannibal Jnr toes at her skull, leans down and pushes the cap gun hard into the soft Arctic white cashmere between her eyes. He reefs the trigger.

Crack!

A cloud of sulphur overwhelms the aroma of fresh mown grass.

Before I can react, he's onto Marshmellow Fluffy Fellow. With a pointed shove, he missiles him down the pants of his four-year-old friend. Laughter. Squeals. A scratching cushion. Screams. Louder screams.

Splash!

Hannibal Jnr attacks from air, ground and sea.

Carob coloured Cocoa swims for her life, dissolving in front of me in the water trough.

Javelin!

Fluffy Rembrandt spears into the paling fence.

Furry textured wall art.

My eyes double in size. My breathing stops. Silence of the bunnies. Silence from me.

Slowly my arms outstretch and I scoop, scrape, de-submerge and secrete the dazed bunnies behind me. If I had a skirt on they'd be cowering under it. As it is, they peek out from behind my ankles.

I'm five foot nine. Hannibal Jnr barely comes up to my belly-button. But his malevolence feels eight foot tall, universally wide.

No. No. No. How could I think that? Bad Anna. Children can't be evil, I scold myself. He's just a boy who hasn't been around animals. Doesn't realise they have feelings. Geez you're quick to judge. His parents probably work hard; he has a grumpy nanny; he isn't getting the right discipline; he's been inadvertently exposed 24 hours a day to Call of Duty; he's been strapped into a car seat for three hours and just needs to let off some energy. He's mistaken the bunnies for soccer balls or pillow pets. Be nice, girl. You're the grown up, these are your guests! He just wants love.

"So…boys…how about you go unpack, and, uh, don't come in here without your parents okay?" I say all matey-mate.

Hannibal Jnr stares at me. For the longest time.

Eyes like gravel.

From his pocket, he withdraws the cap gun, taps it hard to the temples of the other two boys and pistols them through the gate.

I ah – I ah – squat and shock-quiver with the bunnies. Did that really just happen? Can a kid be like that? Can running an eco-farm-stay be like this?

I gather the bunnies in my lap, a Leporidae quilt. Stroke them. So, so, soft. Delicate brain frames. Crispy-thin skeletons braided beneath voluminous cashmere coats. Hug magnets. Velvet eyes. Cuddly. Crushable.

Cash crunch.

How does this work? Their families are paying to be here. Can we discipline a guest's child or do we have to be all nicey-nicey? Can I give him a lecture on animal rights? What's the protocol? Can I enforce rules I didn't even know we needed? Hmmm, the customer is always right, so do I act Saudi-Arabian maid and not see, not speak? How do you bring it up with the parents in the first seven minutes of a seventy-two hour stay without killing the country hospitality vibe? Is this how our simple life is going to be? Do I have to turn myself into the RSPCA for putting our bunnies in danger? How do I live with myself if our bunnies don't?

Money muddies morals.

For a moment. Minutes.

Then I hit upon a solution: our software needs hardware!

A one hour return trip to Bunnings and the ute is loaded with enough combination locks, chains and reinforced padlocks to protect the Reserve Bank. Gerbil Youth, meet lop-eared bunny lock out. I've even splurged on a seven-pin lock with dimple keys just in case Hannibal Jnr is as good at break-ins as bullying things.

I'm chaining the second gate, an improvised Hannibal Lecter Jnr air lock, when I feel his force field approaching. The one cloud in the sky cloaks the sun.

"Mum says I can go in."

"Tell Mum you can if she goes in with you," I say kindly. Warily.

"She's busy."

"You'll have to wait till she can then mate, just get her to let me know and I'll give her the key no problems," I slip the key in my pocket while a cloud, a very, very dark cloud, slides over his face. "So, I'm heading in for a quick lunch, do you want to check with her now? 'Cause otherwise you'll need to wait till the afternoon animal round."

He stares at me for a moment, then the storm lets rip.

It's an awful, anguished, thunderous sound. Like the sound you hear and the pain you feel, when you get shot in the kneecap.

Mrs Lecter, paperback in hand, sprints down the cottage ramp and across the lawn like a leisure-wear pro. Kneels before him. Holds his shoulders.

"She...she..." He points at me. I'm a line-up of one. Guilty face. Dirty hands. He yowls again so loudly and convincingly, it's as though someone has just stomped on his imaginary shot kneecap. My dirty palms begin to sweat and I have the horrible feeling he is going to accuse me of whacking him. I hold my breath waiting for the witch hunt. He is jab, jab, jabbing his finger at me as though he wants to jab me back to Salem, Massachusetts, 1692. "She...she...," here it comes, "...she won't let me in with the bunnies." Yowl!

I've been holding my breath waiting for the accusation, the inter-rogation, the gallows, but now, deep with relief, I suck in a lungful of oxygen. Innocent! "That's right, only with a parent," I say.

"That seems very restrictive." Displeasure fights its way through the Botox handcuffs on Mrs Lecter's face.

"You can have the key if you like," I offer.

She shakes her head, fast.

"It says in the compendium 'afternoon farm activities start at five'. We'll send the kids to your house then."

"We'll meet here is fine," I shake back, the thought of Hannibal at my front door making me go all Clarice, "you can join us you know."

"Can't do, we're renovating our holiday waterfront. Builder's just phoned, asked to go over by another four hundred thousand. Can you believe it? We're already at a mill. Have a conference call to sort it out."

Yowl. Yowl. Hannibal interrupts, his voice in such pain, his eyes obsidian dry.

"Let's get you a treat," she says, dismissing me and taking Hannibal inside for what I suspect might be a serve of two-storey, gold-plated ice cream. She turns back to me briefly, squeezes his hand and says, "don't worry about *her*."

Ouch. In my 'living the dream' plan I never saw myself as a *her*. I envisioned myself more like a crows-footed Bindi Irwin cross Chief Seattle cross Al Gore cross Hugh Fearnsley-Whittingstall cross Cate Blanchett cross kind, enthusiastic, never cross female Friar Tuckish farmer surrounded by sunflowers, butterflies and lambs...all set to a Snow White soundtrack of trilling la-la-la-la-la's. Hmmm. But apparently our first guests, our first potential converts to a gentler, more equitable, more sustainable way of living see me as a *her*, no "o" on the end.

I make it my mission to get to know them better, and maybe they might get to know me too. Maybe we can save the planet together!

Over the next few days the Dads reveal themselves.

"First class isn't all it's cracked up to be, though Emirates is alright," says White Shirt.

"The cruiser's basically moored at the back door but with all the builders on site it's just too hard to get to," laments Whiter Shirt.

"Nabiac doesn't have quite the same ring as the Maldives, Aspen," they agree, "but it's just for a few days and the kids get to have some exercise."

I don't just get to know the guests through their conversations, but through their garbage too. After just two nights their so-large-it-could-be-a-truck bottle recycling bin needs emptying. Clinking, clanking, echo-y receptacles of single malts, Sam Adams Utopia, Grange, Dom. Empty silvers of Temazepan.

I get to know myself through their garbage too. Judgemental. My cask wine bags don't take up nearly as much room.

"No wonder they don't come out of their cottage till 1 pm," says Andrew, hoisting the bottles into the even bigger recycling bin, weary after shepherding their children from 6am, "they're hungover."

"The kids aren't your responsibility all day," I say. "We're a farm-stay, not day care."

The last forty-eight hours have left me feeling less agritourism entrepreneur and more indentured serf, "plus, the rabbits are safe now with the locks. Let them get up and look after their own kids this morning."

"You should see what that boy did to the goats when he didn't think I was looking," Andrew shakes his head. "If he's out there, I'm out there."

But he has his tired Sarcoidosis slump on. The one where he rests his elbows on the table but can't keep his back from sagging.

"You have some lunch, it's my turn," I say.

It's dry as a split-end but I pull on my gumboots. They're my farm shoes of choice: easy on, easy off and high enough so pebbles of goat poo can't flick in, plus the rubber's sturdy enough to bounce off a set of dreaded Death Adder fangs.

It's quiet and I figure the guests are back inside for a late lunch. I can breathe. Enjoy the space that we came for, space that unfolds like colourful crayon strokes north, south, east and west. No buildings, no judgement, no comparisons, just paddocks and scattered gums, sparkling dams, white clouds furry on blue.

Then I hear a peeping.

Part of the farm purchase included an outdoor aviary and two parakeets. Not my thing, caged birds, but to free the hand-raised pair would mean quick death by raptor or slow by starvation. The eastern side of their aviary is protected by a thick-trunked jasmine vine, its white, flowering tendrils snaking in and out of the fine bird mesh providing scented sweetness and shade for the feathered couple.

It's also providing shade for Hannibal Jnr's mum, who stands hunched below the tin roof, the two parakeets mesmerised by the strange human bird in their cage.

She tweets plaintively: "Could you please let me out?"

"I'm so sorry," I gush, "I didn't realise the door was faulty." I add it to my mental list of things to be fixed, it's a list already encyclopaedic in length and in need of its own librarian.

"It's okay," she says, "my son, he accidentally slid the bolt... ran off to lunch. He didn't hear me."

I hear her lies. Catch her eyes. They begin to glisten.

So do mine.

"He's been expelled from two pre-schools. His kindergarten teacher...so many meetings..."

I gently slide back the bolt.

Slide back my bitterness at being just a lowly paid 'her'.

She emerges from the cage, straightening up slowly; vertebrae, secrets, unlocking one by one.

"I love him, but I think he has a problem."

"Hmmm..."

"Maybe its ADHD?" she asks me hopefully.

Lack of empathy. Cruelty to animals. Devious planning. I'm thinking it might be different letters, letters that look more like p.s.y.c.h.o.p.a.t.h.

"Maybe," I say, wondering now if all the empty alcohol bottles were not about her and her husband's lack of responsibility but rather their lack of coping. Should I offer her a serve of reality or a simple turndown service with chocolates on the pillow? Is it my place to even weigh in? No matter how many books and articles I've read, or what my gut feel is spelling out, there's no psych degree on my wall. The moment passes, she walks away leaving me to start diagnosing myself. Prescription: an afternoon restorative of anti-depressant, mindful, meditative weeding.

For the first twenty minutes as I grasp at grass I grapple with the morality of writing down Hannibal Jnr's name now so I can Google him in twenty years to see what horrific crime he has committed. As I pull at plump petty-spurge I'm pulled to tell his parents more about the *silence of the bunnies* incident so they can get themselves some serious help. As I follow the arteries of amaranthus, my brain finally begins to pulse slower. The tiny tears and crackles as roots separate sound like a far-off firework display. I see green veins, lifelines, interconnections. Alleys of chlorophyll run this way and that, they drain my thoughts until not one thought pumps.

Green peace.

That evening at their final campfire, Andrew and I shepherd the boys as the parents ensconce themselves in conversation, inebriation and diversion, finally taking themselves back to their cabins and leaving us with their kids. No goodbye, or do you mind?

I no longer do.

When Hannibal Jnr tries to lance his brother with a burning stick Andrew deftly deflects it. When he tries to sear a tattoo of flaming marshmallow onto his other friend's face I'm a bit slower so need to race to the garden for some aloe vera to cool the burn. When he flicks embers at our elder girl, it's finally time to call it quits.

"Sorry guys, it's a school night for Lucy, bed time now," Andrew says, pouring two buckets of the-party's-over water on the flames.

Thick grey steam gels into the sky, milking out the stars.

"Home time," he nods toward the younger boys. They scurry off without argument; even they know they're not safe in the dark with Hannibal Jnr.

"You too," says Andrew, warm eyes meeting coals.

And Hannibal Jnr stalks back to his cottage, to a family ill-equipped to help him. To a society unprepared for what he may become.

We put our own kids to bed – they seem so much better behaved and postcard wonderful than they did a few days before – and slump onto ours. The smell of ironbark smoke and ash thick in our hair, the encounter layered on our skin.

"Full on huh?"

"Ah, it wasn't so bad, nobody ended up in hospital," Andrew replies.

The fan above the bed clicks around and around.

"But if we're going to make a go of this, if we're going to put up with stuff like that," he says in his management voice, "we're putting our rates up."

My brain and spirit go into battle. I hate transactions, I want to wean off capitalism, detox from dollars, I want to give everything away – but if we under-price ourselves we'll be used and used up in months. I want to make a difference, but I'm not different enough yet to do it.

I turn on my elbow to face him. "I'm with you."

* * *

IN CASE YOU EVER GET INTO A HOT SPOT, THERE ARE ORGANIC BURN balm recipes on page 213.

ONE DAY PROJECTS

I'm tunnelling through the veggie jungle in search of the pile of volcanic rock dust and guano bags required for today's big tree planting mission. The project involves birthing a corridor of 1 200 trees on barren land to create shelter, food and habitat so native beasties will have room to roam. Hopefully, for centuries to come.

For the first 360 seedlings, we had the help of local volunteers from the Landcare group, but since the cheery elves left, and with all the other things needing doing on the farm, the wildlife corridor has been going nowhere fast. The 840 remaining seedlings are starting to wither in their pots and they've been silently screaming at me to save them. Today's the day.

The veggie jungle smells like a smoothie left out in the sun: notes of lime, too sweet pineapple, mint, decay and an overload of hot chlorophyll. Towers of stalks, squishy green footpaths, khaki bridges. It's a metropolis of chaos, a juicy slum of its parts, but my city of hope.

My knowledge of plants is currently limited to what grew around the blue fibro coastal home of my childhood: suburban lawn carpeted by summer-fragrant frangipani; a memorial rhubarb planted atop Tiger our first and only ginger cat; a clump of sugar banana; a yester-day-today-tomorrow tree and a jade bush at the front door that never worked. "Maybe it only works for the Chinese," my Mum told me, referring to the Chinese belief of Jade-bush-induced prosperity. At the backdoor was a nectarine tree I refused to eat from but would swap its tart, peach-sized fruit with the Slovenian neighbours for slices of plastic-wrapped cheese, and with the Tongans on the other side just to hear their delicious laughter.

I don't know my ovate leaf from a pinnate, a rhomboid from a trifoliate. All I know is I want to sprout a bulbous, jolly green giant thumb and grow food for our family and others. In dirt, with no transnational residues, no man-boob making molecules, no cancer spreading spray-ons. The corporate language and science that have given the world DDT, GMO and Imidacloprid will have no license to speak on this turf. No consonants. No vowels. No capital letters or apostrophes.

"SHUT UP!!!" I yell out loud to make the point.

I also feel the need to go Tarzan on supermarket semi-trailers that colonise the other side of the planet via tangled vines of bowsers, refineries, 550 000 deadweight tonne shipping tankers, pipelines, greed, war and suffering on their way to pick up beans and milk thirty minutes to the north of us. Sucking up more fuel they rip baby carrots and baby spinach from their birthplace, repatriating them to big city distribution centres where they're held hostage for days, de-programmed, bound in plastic, barcoded, then resettled in chains at the Woolworths, Coles and Aldis back in our region where the babies, milk and beans lived all along.

I want all this, and I also just want to eat. And one day, I want to eat and share it all guilt-free and laugh deliciously like my old Tongan neighbours. But this garden hasn't seen a mattock from the previous owners, or we new ones, in months. I can't see the food for the weeds, I can't see the soil sprouting the tangle, I can't find the damn fertiliser bags, I can't see the danger.

Killer animals to Australians are a bit like the San Andreas Fault for Californians: you know the threat is there but you just get on with life and ignore it until you get the shudder. Right now I get the shudder. It's a tremble, a physical vibration, an early-warning, involuntary mammalian alarm to prehistoric danger. I haven't even seen it yet. I have one foot in the air stepping over the handle of a shovel.

Except, involuntary shudder, involuntary "Shit!" The handle has eyes. "Shit!"

Eastern Brown snakes aren't really brown at all. They're just like this fawny-grey, age-cracked wooden shovel handle. That now also appears to, oh! Have a tongue!

Flickety-flick. Flickety-flick.

Unlike inert, eye-less, tongue-less shovel handles, a panel of volunteer laboratory mice have actually awarded Brown snakes the title of second most venomous asp in all the lands. They gave the top crown to the Inland Taipan, also representing Australia, and second runner up to the Blue Krait representing South East Asia. But hey, I'm kind of straddling the Snake Pageant Runner Up and I am so not clapping. Neither, any more, are the mice.

Time s-l-o-w-s. I am in freeze frame. Snake suspension.

One leg over it. One leg behind it. Groin right above it. In that moment before crowning, I'm glad I'm not a dwarf.

"If you see a snake, do not move." I repeat weekly to our kids and whenever farmstay guests arrive. "Stay stock still until the snake moves away."

But I obviously wasn't listening to myself because my reaction is to ever-so-ever-so slowly rewind my leg. It's soft focus slow. I am rewinding time. Reeling in my ankle. My calf is now like a boom mike floating above Miss Congeniality. My weight keeps shifting backwards. My leg is returning over the catwalk, trying to look nonchalant as it skims past the judges.

"Look!" I try and mess with the snake's head, "there's Donald Trump!"

Miss Congeniality is momentarily stage struck.

She is holding fire on her neurotoxin talent.

So, I steal my chance at the title. My talent is flight not fight. I am back pre-shovel. I am back at the khaki bridge, reversing over the squishy green footpath, tumbling backward through the towers of stalks.

The shudder gives way, the legs give way. To lawn. To fret. What if it were the kids? What if it was a farmstay guest? Do other companies run businesses that involve snakes? Does McDonalds guarantee no snakes with that? What about movie producers? Banks? Humanus greedus snakus? Could we be sued for a naturally occurring snake occurrence? What if it had launched? What if there's more? I'm in the middle of the aftershock and I can feel the pressure bandage on my brain.

"You right?" asks Andrew, sandwich in hand, looking down at me sprawled on the grass.

"Brown snake! In there!"

"Oh," he keeps chewing, "was talking with Nic the other day." Nic's a paramedic friend of ours. "She says if you ever get fanged, call emergency and tell them a child's been bitten, then they'll send the rescue chopper. If you tell them it's an adult, they'll just send the road ambos."

"You're kidding?" Asp ageism? If there's not already enough discrimination, winners and losers, haves and have nots in the world, now because of my age I'm one of them too? I wince at the irony that I want a rescue chopper if I need one, but that the fuel that rotors it comes from the Middle East, where kids really need rescuing – not from reptiles, but from the sickening power plays of multiple raced and religioned, oil guzzling homo-snakiens...kind of like me, but minus the religion, but, back to the–

"B-r-o-w-n snake! It's still there!"

"You want me to kill it?"

Hmmm...it's illegal in NSW to kill Miss Congeniality (aka Pseudonaja textilis) unless she's directly threatening you. The statistics also prove most people only get fanged when a) they try to pick up said snake or b) try to kill said snake. Plus, I'm a vegetarian who doesn't even kill cockroaches, even the really big ones that have leather leashes and diamante collars and hang out with celebrities in the kitchens of hatted restaurants.

"Kind of."

"Shovel or shotgun?"

"Well…I suppose it didn't actually go me."

"They lay up to 40 eggs."

The thought of the kids barefooting to their death so close to the front door rapidly overrules the rules, bleeds out the bleeding heart. It's a surprisingly quick haemorrhage.

"Rifle."

Andrew heads to the shed to retrieve the recently-acquired-with-the-farm-rifle, which he sat his gun test for, presented to the police station for, got his license for and which, according to NSW gun laws, needs to be secured in a metal box with a padlock and a combination lock, secured to the concrete floor.

When he's undone the lock to the shed and the two locks to get to the gun, he then needs to retrieve the ammunition which, also according to law, needs to be held in another locked box.

These are great laws to stop hotheads blowing holes in their neighbours, toddlers shooting their mothers in the back, prep students being peppered and robbers running off with Rugers, but, case in point, the laws may also give innocent snakes ample and totally legal getaway time. So, fifteen minutes of unlocking and assembling later, Andrew finally enters the arena. I, his biggest fan and now hitman hirer, wait nervously at the jungle stage door.

One minute.

Two minutes.

Five minutes.

There he is.

Finally.

Emerging…

Bang-less.

Sash-less.

Miss Congeniality has slipped off stage. There will be no victory slither walk for either species today, and I'm…relieved. I can feel in the shaky beat of my heart that "Snake Murderer" isn't a title I would wear well, nor would I be able to carry off the red-spattered ball gown of victory. I feel so exposed; my fake tan of mercy has been smudged. I called for the hit, I shed my compassion skin like a change of outfit and asked my husband to do the dirty work.

I think both the snake and I have dodged a bullet, but my newfound live and let live for killer reptiles doesn't mean I have a death wish, that I'd put the snake on the podium above the kids

"We've got to clean this up!" I chest thump at the jungle. "A proper garden! With really, really wide clear paths, with raised beds, with nets to keep the birds out…"

And I see clearly but don't say: a detailed snake metro with round-abouts for red-bellies, cul-de-sacs for copperheads, bridges for blacks and browns, traffic lights for taipans and tigers and a dirty big detour for death adders. There will be no front-fanging in this veggie patch. There will be no death by venom-induced consumption coagulopathy. It will be a dream garden, a tame, pedestrian-friendly supermarket of produce where no one, here or on the other side of the world, needs to die for dinner.

Andrew interrupts my head picture. "Forgetting something? Got to get the trees in first, the seedlings are starting to die." He's right, and he's still got the gun. So, I nod.

In one corner of the farm I want to madly pull things out, and in the other I want to madly put things in.

It sounds easy to plant a tree, you just dig a hole and plonk it in. That's why we thought 1200 trees would be a breeze. But no. To correctly plant a tree, you must first kill the surrounding grass; then scrape and gouge and dig into hard clay to make a hole. Now carefully add a prescribed amount of a secret growing formula which may or may not include volcanic rock dust and guano…if I could just find the bags. Water in the secret formula; then immerse the potted seedling inside a bucket of water where you leave it until the halting of bubbles signifies the rootball is now truly soaked. Now extract the truly soaked seedling from its tube; gently tease out the roots; then plant the seedling, firming down the soil around it. Continue on by wrestling with an environmentally-friendly biodegradable weed mat so it snugs around the stem, covering the bare dirt to suppress nasty, competitive, I'll-have-your-corner-office grass types.

Stretch your back. Forwards. Kink. Backwards. Kink. Ow. Ow. Bend over again.

You then assemble a triangular tree guard – unless someone has done it for you – capable of repelling rapacious wallabies and hangry hares; place the rapacious-wallaby-and-hangry-hare-repelling triangular tree guard around the seedling and then precisely align it on top of the biodegradable weed mat; get a timber stake, get a few splinters, drive the splintery stake through the heart of the triangular tree guard, pick up a mallet:

Whack. Whack, Whack.

It bleeds. You bleed.

Refill the watering can and the bucket from the dam a couple of hundred metres away; Jack and Jill it back. Pour. Walk. Refill. Pour. Fall down the hill. Do it again the next day. And then the next day. And then the next day. You need to do that x 1 200 trees. And if you're doing it with three children in tow, in the paddock furthest from the farmhouse, you also have to continually convene for three-way junior Judge Judy sessions, splinter-removal, food and "isn't this fun?"

I slump to my haunches, body tired, but head a writhing jungle of thoughts: why did we start this planting project when there are bills to be paid, a business to build, a dairy to be fixed, a veggie jungle to be patched and fences to be goat-proofed? Why are we going against the grain of the landholders who one hundred and fifty years ago worked even harder than us – but in reverse – to kill the rapacious wallabies and hangry hares, to fell the virgin forest thus creating a blank canvas to blanket with sheep and cattle. Why are we now breathing life into these wildlife-supporting, canvas-colouring seedlings, four in a row, along nearly a kilometre of our western boundary? Shouldn't we just kick back, watch the box and get takeaway?

Why are my elderly parents, on day release from their retirement village, sitting right now on the deck of the farmstay cottage arthritically assembling hundreds of tree guards? So many tree guards in fact that they are piled to the roof and every hour, if we haven't collected them in time, avalanche back down onto their benevolent silver heads.

I rise. Stretch. The brain jungle recedes behind the wood, behind the trees.

Maybe it's so one day a child will take inspiration from a koala in the wild, eating from a Forest Red Gum we've nurtured. So one day colourful lorikeets will sup nectar from mature Banksias and Bottle-brush before taking refuge from predators amongst the branches of Hakeas. So one day, in their time of need, someone will have access to shade, to timber, to warmth, to the termite-resistant Turpentines for fencing and to the Tallowwoods for nectar for bees.

Maybe it's so one day we will be unable to wrap our arms around trees we once held between fingers.

Maybe it's so one day the snakes will be able to hang out further from the house.

Maybe it's so one day we will feel like we've done our bit and will relax and laugh deliciously.

Whatever the reason, this, and the veggie jungle, are "one day" projects. One day projects I'm beginning to understand, that could quite possibly take our lives.

* * *

YOU'LL NEED A LITTLE MORE HELP IF YOU GET BITTEN BY A SNAKE, BUT IF you're just getting bitten by bugs, I've concocted a great recipe for an itchy bite balm on page 215.

MAKING MISTAKES

I want to know stuff. I want to get things right. I want to know which foraged mushrooms I can serve without killing people, I want to know what herbs will drift me off to sleep, I want to know how to be a contributor not just a consumer. That's why I'm out in the garden as it dews, like right now, and sometimes still out when it's dark.

I want to try everything, share everything, do everything, well. I want to smell and taste and harvest and heal, properly. Voracious for knowledge, for wisdom, for a future, I devour herbal medicine and sustainability books till Rocky rooster's dawn-rocking first crow. I'm a hybrid hippy workaholic, not yet tie-dyed enough to relax.

I want to know who you are. Yes, you! Little bug, little bug way down there in the green stuff.

So many plants and bugs remain enigmas to me. They're like throbbers in a mosh pit, indistinguishable, entwined, a kaleidoscopic blur. Weed or wasabi? Poison or potion? Herb or spice? Beneficial aphid-eating lacewing or munching, destructing winged-grasshopper? The whole first season I'd marvelled at the multiplying orange ladybirds, wishing them well against the nasty little yellow micro-bugs destroying the potato and capsicum leaves.

"Look at all the lady beetles," I had gushed like a proud mother to Philomena, the Italian-heritaged local garden guru when she popped in for her first visit a week ago. Philomena's saintly organic presence was like having Yoda and the Pope jointly consecrating the garden. I toasted in the warmth of her horticultural halo.

"Ah," she sighed profoundly. "That is the twenty-eight-spotted ladybird."

"She's so beautiful, I love them!" I said, feeling so proud to have enabled such a clan.

She looked at me warmly. "This is the only ladybird you do not want."

What? How can this be? "I thought we wanted lady beetles?"

"Yes, just not this species. Too many spots. See how many more spots they have? That's how you can identify them, and the damage. They will just eat through your squash, potatoes, capsicum leaves." My child.

"But don't they eat these little yellow bugs that are eating the leaves?" I had lifted a leaf and shown her the tiny slater-like shapes massing on the remaining potato leaves.

"No, the yellow ones are their babies."

I fingered the iron brown lace of the potato leaves, so dry and desiccated. Who would have thought? A bad ladybug? Of all the five thousand species in the world, I had enabled the worst one to multiply into stormtroopers. And here was PhilomenaYodaPope, the biody-namic Buddha who plants by the moon, grows un-holed leaves and ferments the world's best pickles, a close-up witness to my ignorance. It was like lying on a microscopic slide being dissected by my own inabilities.

My face spotted with the embarrassment of an amateur, the mark-ings of the worst farmer on the planet. Ladybird, ladybird fly away home...and take PhilomenaYodaPope with you so she doesn't notice anything else I've done wrong. Quick, get her all-knowing, wholly benevolent presence out of here. "Let me get the gate for you. Thanks for coming. Bye Philomena."

The top gate is a bit of a focal point for the farm. There are always comings and goings, farmstay guests, randoms, PhilomenaYodaPopes. One coming left the metal-barred gate bent in half when someone mistakenly accelerated into it. I've been asking Andrew to unscrew and tractor back over it to straighten it up, but there's always something, or someone, more urgent, coming or going. So, it sits there, annoyingly, welcoming people – all bent out of shape.

Like today. It's 8 am and I'm running a counterinsurgency against the incumbent twenty-eight-spotted lady beetle battalion when someone starts blaring their horn up at the gate. I straighten my back, pull off my gloves and walk up the driveway. Who does that, I wonder? Who sits in their car in a quiet country setting and blares their horn?

"Can I help you?"

"You can bloody well keep your goats off the road," tirades a red-nosed, stringy-haired sixty-year-old I've never seen before. "You gotta dead one out there. Could have killed someone with your shitty farming and shitty fences."

Um, good morning. Um, that was a little abusive. Um, I think I want to cry.

Tyres spinning, he reverses, shunting up dust like he's just shunted up me.

I fumble with the gate latch, run the fifty metres of dirt to the bitumen. Ohmygod, not one of the goats. Ohmygod, how did it get out? Ohmygod we are such shitty farmers. Such shitty humans.

Two thin lanes of bitumen liquorice bound the north side of the farm, running from Nabiac in the east to Gloucester in the west. The liquorice curves dangerously as it arrives at the eastern tip of the farm, which is where you can continue on to Gloucester or skid left into the cinnamon-dirt road down to our entry. It's where the liquorice and cinnamon meet I see the fur. Creamy, bloody, like a pile of ripped up carpet after a serious Shiraz spill. It's not one of the babies, too big for that. It's one of the adult goats. Oh no, not Princess? She's big, fawny. But it's so hard to tell because she's now a carpet tile and I don't want to get too close. I want to know but I can't bring myself to see.

31

"Andrew! Andrew! Andrew!" I blare my horn, all bent out of shape. Imagining the car running into our goat. Blare. Imagining the mistake we have made causing so much pain. Blare. Imagining Mr red-nosed, stringy-haired sixty-year-old – whoever he was – coming back for another go at my shitty little farmer self. Blare.

"What's up?"

"One of our goats…got hit by a car."

"I just fixed the fences."

"A man, he was really angry."

"Who is it?"

"I don't know who he was!"

"I mean, which goat?"

Andrew has his favourite goats. There's Sarah with her man-beard, our LGBQTIA goat, who would be by his side, or at Mardi Gras, all day if he let her. There's Tracey of the massive udders, so massive the left one, when full of milk, bounces along the ground like a full airship. Andrew had felt so sorry for her he bought a human maternity bra and spent hours fashioning it with luggage straps to form a patent-worthy piece of milk-filled, airship-lifting, lingerie-engineering he christened "The wUdderbra". Then there's Princess, a totally chilled out caprine, born on our elder daughter's birthday, hope it's not her either, that would be hard to explain…there are about twenty-eight goats all up and he converses daily with each and every one.

We slow down as we approach the rouge shag pile at the crossroad. Andrew moves ahead. Lowers to his haunches.

Tick. Tock. Tick. Tock.

He must be feeling really, really bad.

"Anna," he says, and I await the name, the guilt. "It's deer."

"Who? Dear?"

"Yes, dear."

"The sheep?" Dear is one of my favourite sheep, friendly and wail tagging and forever in desire of a chin scratch.

"It's not Dear…it's a d-e-e-r!"

"It's not Dear? It's not a goat?"

"Look at it."

And then I see my mistake: the different snout, the slender limbs, the fawny colour, not of Princess or Dear, but of a young feral doe. This time the looking, not avoiding, brings brief relief. It wasn't one of ours, it wasn't our fault after all – but I still feel it could have been given the state of the old fences, the nature of goats and the tendency for guests to leave gates unlocked even though on each gate we have a little octagonal reminder saying, 'Shut the gate mate'.

I get the shovel, Andrew gets the wheelbarrow. Can't leave it here to cause another incident or accident. But car after car goes by slowing down, gawking, thinking we've let one of our animals die, that we're a farming menace to their Landcruisers, their HiLuxes, their Pathfinders and Patrols.

Pry it up.

Stuff you Mr Horn Blarer and passing traffic as you go around for the rest of your life thinking we're shitty farmers with shitty fences.

Scrape it up.

Stuff you Mr Horn Blarer and gawking passengers for reminding me how much I hate people I don't even know, thinking they know me.

Shift and lift it up.

Stuff you Mr Horn Blarer for making me think I made another mistake. And thank goodness you didn't see it when the Shetlands got out!

The wheelbarrow is now full of rouge shagpile.

"I'm going for a shower before the guests get here," I say, flushed with exertion and roadside embarrassment. I'm going to wash Mr Horn Blarer – and the dust and the blood –– right out of my hair. And I'm going to treat myself: no apple cider vinegar hair rinse today, I need some luxury to coat the embarrassment. Today I'm going to indulge and try out my first ever whole egg conditioner.

Fresh free range egg from aisle two of the farm thanks. Crack. Whisk. Apply to hair. Sing a few tunes. "You gotta fight, for your right, to parrrrteee!" Rinse. "I get knocked down, but I get up again, you're never gonna keep me down..." Rinse. "Somewhere...over the rainbow..." Rinse... Oh, this hot water is so relaxing.

What the?

The dust has rinsed, the blood has rinsed, but the egg – like Mr Horn Blarer – has stayed. Cooked. Scrambled. Set into tiny forever omelettes in my locks. Egghead! Why didn't I pay attention to the book when it said, *rinse hair with cold water so the egg doesn't cook...?*

"Mum?" calls Lucy.

"In-the-showerrrrr!"

"The guests are at the door."

"Can't you get Daddy?"

"He already checked them in, but he's gone back out to bury the deer, can you come out?"

Out! Out! Get the egg out. That's what I'm trying to do!

"Mum?"

"Arghghgh! Won't be a sec."

I emerge, hair whisked with albumen, a full head follicle meringue. I ponytail, then wet bun it to minimise the omelette look, shove a cap hard over the top, add a bit of salt 'n' pepper. Time to meet more people, meet more impossible-to-meet expectations for the first time.

The first thing I notice is her luscious black hair. So black, so sleek, so Cleopatra. So opposite to egg white. The next thing I notice is she's so regal, so monolithic, so Sphinx.

"I think you've booked us into the wrong cottage," she tuts, the Tutankhamun of tuts. "We asked for a cot." Tut. Tut. Tut.

"Sorry about that, got a bit caught up this morning, will bring it down straight away."

"Hmmph." Hieroglyphic expletive.

Shitty gardener. Shitty farmer. Shitty hotelier. That's me, the one with holes in her potato leaves. That's me, the one who failed the roadkill species degree. That's me, the pyramid builder, lugging the overdue port-a-cot down to Cleo.

Yep...that's me...the one with egg dangling all over her face.

This just isn't the farm, career or village where you can hide your mistakes. Everyone sees, everyone knows. And now with the Internet, it's all over TripAdvisor and Facebook too. Mistakes are magnified... not so magnifico.

At the barbeque shelter I stop for breath, for composure. Dirt-floored, it's backed with trellis to protect from the westerly sun. Two timber poles at the front hold up the roof and sheltered beneath are toys for the guests: a pedal tractor, balls of all shapes and sizes, a cricket set, my feelings. I slide onto the timber bench, not quite ready to face the wrath of Cleo. It's dusking now so the automatic lights turn on.

Right in front of me, a shiny, chunky bug wings its way whack-bang into the pole. Whoomp! A bug! In an accident! I'm hit by the shock of it. Surely he saw the pole? Surely he knows how to fly? Surely he knows you need to concentrate when flying? Surely bugs know how to be bugs? Surely bugs don't make mistakes? Especially in mid-air? In their domain?

But it did. I'm pole-axed by the absurdity of it.

I look to where he's fallen: it's a scarab beetle, one of the thirty thousand species world-wide, but one of only twenty-one species endemic to New South Wales. It's the one that flies during summer, herding us into Christmas. It chases the light, but now it rolls in the dust. I can't quite tell if it's a Washerwoman (Anoplognathus porosus) or the rarer – of course – King (Anoplognathus viridiaeneus), but I'll side with the royalty this time.

Staggery, he's shaking it off, back on his six teensy feet, shuddering then stretching his stubby wings, ready for take-off again. I wonder if he's embarrassed, defensive? I was watching after all. But no hint of that, he's up and away, a little bit ziggy, a little bit zaggy, but pole-avoiding this time. Concussed, but in flight after his inadvertent mishap.

It seeps in through the egg nog: if even all-natural, since-the-beginning-of-time bugs can make mistakes and shake it off, why can't I?

What an inspiring bug!

On the farm stereo I can hear the safe bleating of goats, the gentle baa of Dear, and in my own head, the fading call to fly perfectly.

Makes me want to take my cap off, let my omelette see daylight. Fail out loud.

Makes me want to invite PhilomenaYodaPope back for another visit. Learn out loud.

Makes me want to track down that little scarab of a beetle and Scaramouche, Scaramouche, do the fandango.

Live out loud.

* * *

LEARNING NOT TO FUSS? TRY THIS FUSS-FREE CHUTNEY ON PAGE 247

DON'T MENTION THE GERMANS

$\mathscr{M}$y dirt encrusted fingers, fresh from an assault on a microscopic corner of the veggie jungle, swoop the vibrating yellow handset from its office cradle. It's a great farm phone, tough enough to drop, fluoro enough to find again, and holds a splash-proof signal for 200 metres around the house so you can walk and talk as you herd and haul. This farm is multi-task central. It's the biggest place I've ever lived in, but with so many bills to pay there never seems to be the room or time to sit and natter.

"Vee see your farm on the Internet. Vee voof? Start tomorrow?" says Miss Germany.

Voof?

"You'd like to make a booking for the farmstay?" I ask. I'm getting better at Asian accents, but Europeans get me all the time.

"Nein! Nein! Voof!"

Voof?

Ah! WWOOF!

Simple living magazines form the complicated pillars of our bookcase. They're full of stories of Wwoofers – aka Willing Workers on Organic Farms – who travel the world swapping help for food and board. We've never entertained the thought of being a host, I mean, who would want to invite strangers into their frenetic fishbowl of family life? Who would want their extremely soiled laundry on show at the dinner table? And what sane person on this spinning globe would choose to cook for extra people at the end of a sixteen-hour day?

Hmmm…maybe people with rampaging veggie jungles. Maybe people with 1 500 trees to plant. Maybe people who dream of saving the planet in between washing twelve sets of guest linen, cleaning two cottages, trimming alpaca hooves and hauling honey buckets. Maybe people who, when they finally crawl into bed at 1am, have recurring dreams of Oompa loompas arriving at the farm offering help, chocolate and a lifetime supply of quirky tunes.

"We've not thought of hosting backpackers before…" I say, eyeing a throat high stack of bills and three gut-churning 'to get to' paperwork piles titled: 'Farm Plan', 'Life Goals', 'Urgent Maintenance'. Just last night Andrew had said, "we probably should have just called this farm 'To Get To'." It was a little too much of a self-fulfilling prophecy for me, naming the farm would have to wait too.

"Hallo?" The German checks back in, not sure I'm still on the line. Her accent evokes governess Maria from the Sound of Music, which makes me think of friendly Alice the housekeeper from The Brady Bunch, which makes me think of oh so helpful Albert from Batman, which makes me think of hairy-legged Mrs Doubtfire. Imagine… imagine, if they came to live with us? Imagine what we could get to between our child labour, fictional TV servants and Oompa loompas! I'm thinking this phone call might be the golden ticket to getting things done.

"How long do you want to stay?"

"Vun veek."

One week. An irresistible seven days of help. "Vunderbar! The Greyhound Bus has a drop off just down the road, I'll get you from there."

And just like that we're getting an Alice and Maria! Actually, I don't know that, I didn't ask if they were two girls, or a girl and a guy. Maybe we're getting an Albert. Or a Mr/Mrs Doubtfire. Or multiple-birthed, orange-faced, slightly ornery Oompa loompas. Don't ever give me a job in HR okay!

"Where are they going to sleep Mum?" asks Lucy, shutting the door of her room behind her.

"I suppose there aren't any guests in Cottage Two for a couple of weeks, how about down there?"

She's happy with that and skips up the corridor. "We're getting some Germans! We're getting some Germans!"

I start feeling uneasy. It's the inviting strangers into your home thing. It's the extremely limited security check I did thing. And it's the 'will they like us?' thing. The whole WWOOF system is based on recommendations, if you're a good host, you get feted, but if you're a bad host you get flamed publically, on the Internet, for your neighbours to see, for your competitors to see, for other potential Wwoofers to see. Through two Germans I've just invited a whole world of strangers into our life and lounge room for a potential Trip Advisor-like humiliation of our farming and family.

But they're coming tomorrow, and Andrew's coming now.

"Guess what Daddy?" says Lucy, "we're getting some Germans tomorrow!"

"German shepherds? German cars?"

"Wwoofers babe, free help for a week!" I enthusiastically sell.

He looks at both of us like we're speaking a different language, but he's used to interpreting, so bears with us.

"You know, Wwoofers! Backpackers! Helpers! It'll be great," I say, "they can help me in the garden, then help you fix the irrigation."

"Help me polish the saddles," sells Lucy.

Wary Jack eyes us all warily. "Well I don't want one," he says.

"That's okay cause we're getting two!" says Lucy.

"Worth a try I suppose," Andrew says, only barely agreeing to our high-pressure sales pitch. "There's a lot to get done."

"Imagine if it works! If they give us a good review we could have help all the time and the kids could meet people from other countries. We never leave this place so we can bring the world to them!"

"Germans huh?"

"What's a Derman?" asks Rosie.

"I suppose we'll find out tomorrow," winks Andrew.

Tomorrow comes around quicker than a German engineered sunrise. I sit in the Caltex carpark awaiting the bus. The Pacific Highway, 960 km long and linking Sydney with Brisbane, thunders to the sound of Kenworths and Macks, Internationals and Sterlings. Passenger cars scuttle like ants avoiding boots. I hear the Greyhound before I see it, the steady deceleration, the lowering of gears, ants parting for it left and right. The door wheezes open with a hydraulic hiss.

"Got some backpackers for you love," says the driver coming down the steps. He opens the underbelly of the bus, dragging out two backpacks, with shoes tied on, with sleeping bags tied on, with hats tied on, with water bottles tied on...and I'm hoping no bedbugs tied on. "Have fun with those," he says. "Have fun with them."

Did he just roll his eyes?

"Thanks," I say. I have two booby-trapped packs. No backs. Finally, they appear on the stairs. A she and a she. Blinking, blank, disorientated; like they're arriving at a resettlement camp.

"I'm Anna, welcome!" No response. "Wilkommen!!!" "What time did you leave Sydney?"

"Five."

"Oh, you must be so tired, let me get your bags."

And they let me. I let me. It's a mix of country hospitality and please-like-me.

I hoist the maroon one onto my back, the shoes and bottles jangling all about. Then I bend down for the navy one. Okay, this is heavy, but I get frontal lift off, I have a water bottle wedged into my abdomen and Nikes gnawing into my navel but it's only 20 metres to the car. 19. 18. 17.

"Have you eaten?" 16. 15.

"Nein." 14. 13.

"Would you like some lunch?" 12. 11. 10.

"Ja." 9.

"Okay, we'll head straight to the farm then," 8. 7. 6. "Get you settled in." 5. 4. 3.

2.

1.

My torso explodes with relief as I offload their mobile homes into the boot.

They settle into the back seat while I pant and check them out in the rear vision mirror: early twenties, brown straight hair, white straight faces, matching expressions. She sour. She dour.

Must be exhaustion. Must be overwhelmed. I try and channel benevolence: I'm my Dad visiting prisoners in jail, I'm Gandhi visiting the British, I'm Morgan Freeman in Driving Miss Daisy. We arrive at the farm and I even oomph their bags up the ramp to their cottage while they take in the view.

"Here you go guys, the whole cottage to yourselves. Just come up to the house when you're ready for lunch."

"Vee come now."

It's only seventy-five metres but they get back in the car so I Morgan them to the house.

"Right, so here's the kitchen, feel free to help yourselves," I throw open cupboards, fridge, breadbox, freezer. "We don't bin any food scraps, just put them in either the chicken bucket or the compost bucket, everything needs to go back in to enrich the soil. No chicken in the chicken bucket though okay? We don't want our chickens going cannibal."

Awkward pause.

"Right, well, when you've had lunch, just meet me in the garden. It's just there." I point out through the kitchen window to the veggie jungle. "See you later."

And I do. Much later. It's about dusk when they appear.

"Vee go to cottage now, very tired. Vee come back for dinner," says Sour.

"Oh, ok." I try and hide my disappointment. I had planned a welcome tour and an inspirational Mandela-Obama-Tony Robbins-Dr Wayne Dyer-Moses-Martin Luther King-Willy Wonka-Steve

Irwin type speech about how we should come together this week to achieve the feeding of the animals, the taming of the veggie jungle and the polishing of the saddles. I want them to love this week. Love us. Help us. But I suppose it too will have to wait.

Hmmm, what to make for dinner? I'm thinking something from my generic genetic repertoire until I work out what the Germans like, plus the kids will wail if it involves any mixing of ingredients – if they could section their plates with the old Berlin Wall they would; only one ingredient allowed per side, set the dogs on any other sucker.

I pull back the glass sliding door to the kitchen. That's when I realise the Germans won't be hard to please in the eating stakes, it seems they like everything.

Scattered across the kitchen bench and table are two empty cans of baked beans, the Dresden shell of a cornflakes box, a dregged 2 litre milk carton, a vacant Nutella, a parched 3 litre orange juice, a liberated tray of seaweed rice crackers. Oh, and there's an evaporated pumpkin soup tin and a barren butter container.

Shoved in the bin are wrapper ruins from a loaf of bread, a 6-pack of crumpets and they've shredded through a pack of cheese. Oh, and there's a mesh bag that once encircled oranges, another with MIA tomatoes and a razed macadamia nut box. At the bottom...what's that? Shiny purple? Shiny purple that had protected the last three pieces of my Cadbury bar, three blocks of so-bad-it's-good, processed milky brown sugar I'd successfully kept safe from the kids, Andrew and myself for nine days, camouflaged inside a brown paper mushroom bag.

I turn to my left, the chicken scrap bucket overflows with four cracked eggshells, mushroom stubs, onion skins, orange peels, banana skins, two apple cores, tomato tops and a lettuce core. Two dirty frypans are trapped on the hotplate, next to a saucepan suffering serious baked beans burns. Battlements near the sink are made up of nine plates, five bowls, two cutting boards and what looks like our entire cutlery set. Nothing's rinsed. It's like voracious veggie jungle warfare has sprung up indoors, chaos quietly vining inside while I've been out taming the mothership. It's Oompa loompas gone wild, Maria and Mrs Doubtfire on crack.

The war on food takes a long while to sink in. I am experiencing the vocabularly.com definition of bewilderment: *'It means not under-standing, but it goes beyond that – it implies a state of complete mystifica-tion. People experience bewilderment when they are utterly baffled by the situation at hand.'*

I go to my non-conflict fall-back position, the practically lying down-slightly bent over backwards-United Nations one: the benefit of the doubt. They must have just taken some of the food for snacks for tomorrow. I bet they'll be back any minute to help clean up. They wouldn't leave it like this, would they? They must have been famished. Starving! Oh so tired. Am I expecting too much of Germans because of the reputation of BMW and Mercedes? What are they expecting of us? Do I say something to them? Is this normal in the WWOOF world? Is this normal anywhere? How will I deal with this for the rest of the week? This sucks. I don't like this feeling. Ohmygod, I think this is how I used to leave my parents' kitchen! Is this cuisine karma? Universal payback for past pantry plunders?

I clean up.

Making dinner I clank pans, gouge onions, skin potatoes and chainsaw out bad bits.

Andrew appears smiling at the door, toeing his boots off, the kids in a line beside him do the same, wobbling and falling into each other. The kids and their farmer father are hungry and tired after another day out planting trees in the wildlife corridor. Sour and Dour seem hungry too, arriving with perfect dining timing.

"Hi, I'm Andrew, you must be the Wwoofers." He shakes their hands happily, not knowing their chequered arrival history. "How was the bus trip?"

"Long", says Dour, like her face.

"Are you a Derman?" asks Rosie with the wonder and anticipation only four year olds possess, like a Derman might be some kind of mythical, magical creature.

Sour nods with flat-lined eyes and mouth, instantly flunking the four-year-old 'are you possibly a fairy or mermaid?' test.

"How long are you in Australia for?" asks Andrew.

Shrug one.

"What do you girls like to do back home?"

Shrug two.

"Where do you head to after here?"

Shrug three.

Soon we'll have a whole wardrobe of German-made shrugs and will need to bag them up for repatriation.

"Andrew's accent can be hard to get sometimes," I apologise for Andrew's quick speaking ways, "Can you understand okay?"

"Ja."

Andrew stops smiling. He's out.

If you've ever eaten dinner in a mortuary, I now know how you feel. No banter. Solemn awkwardness. Grim chews. Sour and Dour could suck the life out of a humpback whale, but to give them credit, they could probably eat one too.

"We go now to cottage," says Sour getting up from the table. Dour slouches out behind her.

"Tomorrow, 8.30am then," says Andrew, firm but friendly as they head out the door.

Andrew raises his eyebrows, "That was a little...stilted."

"How come they don't have to take their plates out?" asks Lucy.

"I knew I didn't want a German," says Jack.

"Well they're here for a week to help and they're tired, so let's be nice and help them out," say I.

"Isn't that why they're here?" asks Andrew, "to help us?"

I go to bed thinking about that, remaining mostly sleepless thanks to the noisy, moving demarcation lines between fair exchange. Toss. Exploiting. Turn. Being exploited. Toss. The desire to get things done. Turn. So I can get to more. Toss. The need to be liked. Turn. In real life and online. Toss. Oh look, it's morning already. Must look into Valium some time.

8.30 am. 9.30 am. 10 am. Sour and Dour finally saunter in for breakfast, German efficiency in question, the brand value of Merc and BMW plummeting globally.

"Right, so when I say 8.30 am, that's actually what I mean." Andrew welcomes them. He stands firm at the door with two long, handled shovels, "today we're going to dig two holes."

"So, grab some breakfast first," I say over-brightly, trying to jolly up Andrew's calm directness. And then to Andrew, "babe, there's a message in the office you need to respond to." I shadow him out.

"Far out, you can't speak to them like that: 'Dig two holes'. They're going to think you mean for THEM! They're going to think you're the backpacker killer and they have to bury themselves!"

Silence.

"I doubt they've seen Wolf Creek," he finally says, referring to the horror movie franchise spawned by the real life 1990's discovery of seven young travellers murdered by Ivan Milat in the Belanglo State Forest. "And there's just no use them being here if they're not going to help."

"Seriously Andrew, they're out here in the middle of nowhere, maybe that's why they're... acting weird – they're scared! You really, really can't talk like that, you need to give them more information, like, "today we're going to dig two holes – FOR irrigation."

He grunts. "If they were scared they would have taken their plates out."

"But they might review us and say you were mean."

"Well I'm not, I'm just on time."

Argh! He's right, but I'd agreed to a week and I feel like I owed them that, and what if we tried to move them on? Might they turn Wolf Creek on us? What if their backpacks are stuffed full of malice and malevolence as well as half our pantry? What kind of review would they write? So, over the next six days, with lots of reassuring, cheerleading and spoon-feeding, they eventually do start to help.

They perfect the two-finger weed pull, like one would for valuable Saffron stems. Thumb and forefinger close preciously around a blade of grass. Rests there. Then a gentle pull. A sigh. This is the hardest work in the world. Sweat breaks out on the brow. They need water.

More water. More food. More food. Another break. Ice. Another break. A blade of grass. By sunset they have each accumulated a pile of weeds big enough to fill a tea cup.

We move them on to the saddles. They perfect something we have never seen before, a barely visible to the naked eye arching of the finger, a swirl of cloth glossing beneath microscopic, crescent-like movements. Over the course of two days of beautiful-to-behold surgical finger-ballet – interspersed with copious banquets – 1/200th of one saddle gleams. Lucy, way less than half their age and height, completes the other two saddles, the bridles, the halters and the remaining 199th for them while they digest in hammocks.

On day five and six they finally dig their own graves with Andrew, perfecting such a slow and precise archaeological excavation of the irrigation trenches that, if the tomb of Queen Nefertiti is ever discovered, he will recommend them to Egypt's Supreme Council of Antiquuities to assist with the delicate unearthing.

"How much longer are they here for?" he asks exasperated.

"Last supper."

"Amen."

Dinner is slices of German surly baked with no eye contact and no spice. There is just no jus. At the end of the meal they leave for the cottage. Finally, conversation breaks out.

"What is up with them?" asks Lucy. "They're adults and they can't even polish a saddle. They don't even smile."

"I don't like Dermans," says Rosie.

"I told you," says Jack.

"It's not all Germans guys," I murmur like a side-lined Kofi Annan. "It's just these two...sorry they weren't more fun."

"I'm sorry they weren't more help," says Andrew.

His honesty is the liberation we all need, the kids and I laugh raucously, so very un-mortuary.

"So what if they give us a stinking review," I say, "I never want to go through that again anyway. You guys?"

Everyone nods in agreement.

I wake fresh the next morning, almost doing an overhead throw of

Sour and Dour's booby-trapped mobile homes into the boot of the car. There might not be any food left in the kitchen, but I can sure taste freedom and redemption as I drive them to the Greyhound.

"Have fun on your travels," I say as they get on board as sourly and dourly as they first got off. "Have fun with them," I say to the bus driver, passing on the knowing eye roll that will surely backpack around the nation with them.

I practise eye rolls in the car on the way home. Widen the eyes. Up. Right. Arc round low to the left. Up again. Wider. Try not to break into a smile at the end. Circumnavigation of optics feels surprisingly good – a bit like someone who no longer worries about being liked.

* * *

YOUR EYES WILL ROLL BACK IN YOUR HEAD WHEN YOU TASTE THIS delicious and easy Reibekuchen (German potato pancake) recipe see page 227.

FARMERCEUTICALS

*G*lass bottles bulging with goat milk soldier the fridge, forcing out all other occupants. Buckets of sticky beeswax glue the shed doors closed. Jars of honey bottleneck every spare shelf. Our menagerie is producing but we haven't quite worked out what to do with it all.

"We could host a world record attempt for the most raw milk chugged in a session," I say to Andrew. Next.

"Donate the wax to bees in drought-afflicted areas?" Next.

"Find some modern-day Cleopatra's, an Olympic swimming pool and go a communal milk and honey bath?" Next.

What to do with all these raw materials that are pumping out? I'm overwhelmed because I was the kid at school who couldn't sew a straight stitch, couldn't keep the clay on the pottery wheel and cheated a German cooking class in eighth grade by reheating frozen Pommes Noisettes and calling them 'Kartoffelpuffers'. I just didn't get the craft or home-making gene and would throw out a shirt if it had a button missing. But we're on this journey to be contributors – not consumers – so I'm going to have to learn.

I actually want to learn; nearly four decades into life I'm sick of being useless and in need of huge supply chains to exist. I want to fight the heritage of growing up with Woolies and Kmart and never needing to make anything, grow anything, fix anything or really do anything. I want to fight the government-induced coma of consumption, because I can see the only production they've inspired in me in decades is to produce my credit card. And more and more I wonder what's the point of economic growth when the economics of happiness and the long term regenerative value of the environment is bankrupted?

So that's why I'm going to teach myself to be useful, I'm going to make things. And if they're any good, I'm going to sell them too so I can make a living from kinder living.

I start by going through the bathroom cupboards. Hmmm, what do we spend money on as a family, that I have the raw materials to make?

Yikes, there's that skin cream that cost me a fortune...but the actresses' skin looked so lovely in the ads! There are lip balms and creams for Rosie's eczema, hand moisturisers and all types of soaps and shampoos. I hit the internet and start researching, gob-smacked by the synthetic ingredients in the commercial products we've been slathering on our skin, and basically eating through our pores: Triclosan, PEG-150, Disodium EDTA, Propylparaben, Oxybenzone. Yum yum, delicious, I'll have a plate of each and a side of endocrine disruptor and formaldehyde releaser for dessert thanks.

It's amazing how a sense of outrage can propel one into becoming a mad scientist...or a sorceress. For three weeks I bubble, bubble, toil and trouble over vials of essential oils, bubbling cauldrons of butters, pans of molten wax. One minute the house smells of coconut and the tropics, the next of lavender and Nonna's hankie drawer, the next of orange orchards and Thai gardens. Olive groves sprout in the lounge room, peppermint runs rampant in the living room and I can't tell if the fresh breeze of tea tree comes from outside or in.

I add a bit more honey to that one. A pinch more beeswax to this one. Another slurp of goat milk here, more cocoa butter there, a drop of vanilla, then another. Eventually I have five different lip balms, four different soaps and six different moisturisers. I package them up, send different sample to friends with accompanying scoresheets, and await their feedback.

I also get some from Andrew when he goes to make an omelette. "What's this stuff in the whisk? What's this stuff in the pan? What's this stuff on the floor?"

In my creative frenzy, I've managed to apply beeswax to nearly every pot, utensil and surface we own, so for the next five days I go through three rolls of not-very-environmentally-friendly paper towel, two razor blades and two flagon-sized bottles of eucalyptus oil trying to remove it all. When the kids come into the house from school, the fumes of eucalyptus make their eyes water, the lining of their nasal passages fry and they ask to eat outside.

Just as I get the place clean, the feedback starts rolling in. I gather all the paperwork together and sit down to go through it. I can't help but envision Cate Blanchett advertising my organic moisturiser, my natural lip balm protecting Angelina's pout, and bars of my goat milk soap in demand at home and abroad...but I'll settle for a table at next month's Nabiac Farmers Market too, I've already reserved it for my big debut.

I bring myself back to the feedback, which is neither cosmetic or skin deep, but at the molecular level.

"Too grainy, too runny, too harsh."

"Too soft, too smelly, too greasy."

"Too dry, too gooey, too gross."

Fail. Fail. Fail. Fail. Fail. Fail!

"You okay Mummy?" asks Rosie.

"I'm fine, just the eucalyptus – bit burny in my eyes."

"My skin is burny too," she says, showing me the red of the eczema behind her knees, and the blood of fresh scratch marks.

"Have you been putting on the chemist cream and washing with their soap?" I ask to her nods. "Would you like me to try and make something for you to help?" She nods again, goes back to scratching.

Over the next few days I research even harder, mix even faster, buy three more whisks. I use test tubes and eyedroppers and write down every measure of each batch, every temperature, every method, so I can re-make it if something eventually works.

"You realise you never let us buy anything tested on animals," comments Andrew as I rub the latest batch into his skin. "But you're actually testing on me and the kids."

"And not one guinea pig has died yet," I laugh nervously, hoping nothing I'm doing will flare Rosie's eczema more and send her down the steroid path.

Eventually they're happy and so I am. The fridge is empty of its milk; litres and litres converted into creamy cakes of mild soap. The wax is out of the buckets, gently mixed and melted to fill tubes and pots. The honey jars clear the shelves, finding a new home in the boot of the car. My goodies and I are off to Nabiac Farmers Market for our virgin outing, and even though I've rehearsed in the privacy of our shed how I'll set up my table, I have flutters in my stomach like I'm heading off on a blind date.

Nervous I'll be late, I get there so early, the doors to the showground hall aren't even open yet. I sit in the car and slow cook in my insecurities. What if no one talks to me? What if no one likes the samples? Have I priced things right? Will they like the honey? Will I even be able to pay the stall fee at the end of the day?

I see the lights flicker on inside and go to meet Helen the organiser.

"You'll be over here," she says brightly, showing me to the corner near the kitchen. "Just keep your table back a bit so there's room for Lizzie the bread lady."

Wow, I think, for the first time I'm going to be on the other side of the table, like a real producer. On my fifteenth trip back to the car to get more boxes I see the gingerbread lady. Petra is the pied piper of the village, a walking candy store smelling of nutmeg, sugar and fresh out of the oven warmth. She's been a marketeer for years and her husband Rolf custom made the shelving for her stall.

"Anna," she commands my attention. "This will not work for you, you need boxes like these," she points to her hard plastic containers, "else things will be crushed, get wet. You need to be able to stack them."

"She'll work it out," says Helen. And I wonder if I will. There's so much to learn and as the hustle and bustle builds, everyone seems to know what they're doing but me.

Slowly I build my display. The honey jars out first; five hundred grams per bottle of summer in liquid form. I put a taste tester at the front and next to it, a cup of wooden paddlepop sticks people can use for dipping. Next are the beeswax lip balms: pots of spearmint, pots of vanilla and heavy duty tubes for those with cracked lips. Next to them I line up jars of Farm Balm, the all-round moisturiser that's healed Rosie's skin and feels so good on my face too. Across the front of the balms I lean a hand-written sign: *Farmerceuticals*. At the far end of the table I spread out baskets of goat's milk soap: plain bars, bars with organic lavender, bars with honey and bars with lemongrass, lime and orange.

The first customers begin to wander in from 8 am. At 11 am I make my first sale... to Helen. Then someone buys a lip balm, someone buys some honey, someone buys some soap. Someone buys some more honey. More honey. Someone buys a Farm Balm.

By midday, pack up time, I've made $68. But more importantly to me, I've finally made something by myself – and of myself – and I'll never buy a commercial moisturiser again.

* * *

I AM SO HAPPY TO SHARE THE ORIGINAL FARM BALM RECIPE WITH YOU and yours, it's on page 219.

LIFE'S BULL

The bellows of Fun Factory the cow can be heard across the valley. It's the sound of a rather large, hairy woman in labour. No gas. No pethidine. Kind of like the guttural groans I heard from the woman going natural in the bed next door, while I lightly laboured with Jack on floats of gas, epidurals, Enrique Iglesias tunes and hospital-issue Lemonade Icy Poles.

I grab my camera and run.

Miniature Galloways are the cattle breed you choose when you want Panda Bears in your paddocks. They are double-fluffed, white-bodied, black-eared, black-nosed, round barrels of cuddles; framed by thick, just can't fake it, love-me eyelashes.

"I don't know...they're definitely a girl's type of cow," said Andrew, "when we first scrolled through images on the Internet.

"The Dexter's are boring," I countered, "and the Friesians would just step over the fences. Plus, it means girls like me will buy them from us!"

We need cows to keep the grass down. We need cows to turn sunshine and grass into manure for the veggie jungle. We need another income stream. We need a heritage breed, small enough so it doesn't need grain to survive. We need cuteness in our paddocks!

"These are perfect," I say, smitten by their panda-ness. "No one will ever sell them to an abattoir; they're too cute to eat! They're like labradoodle panda cows! Please! Please!"

Four weeks later Fun Factory, Kiara and Mudcake appeared, freshly blow-dried, so bright in their whiteness and fluffiness it's as though four-legged, rinsed and spun angels have arrived at the farm. When they walked down the ramp, my heart danced furry pirouettes in my throat.

And now we are about to have our first calf, an IVF baby, a lucky lottery, hot-pink-scratchie-ticket baby thanks to a man in a van.

No bull, there's a company called "No Bull". It's for when you are sire-less, i.e. lacking in cattle gonads. The No Bull man sells you big squares of silver scratchie tickets. You stick these on the hairy backs of the would-be-mums, just above the tail. When the cow goes on heat and wants to get lucky, her raging hormones draw the other females to mount her from behind. In doing so they eventually rub off the silver to reveal the hot pink scratchie ticket that can be seen by farmers from far over the other side of the paddock.

I am the farmer far over the other side of the paddock. I see it! We have won the lottery! Fun Factory is hot! Fun Factory is pink! That means Fun Factory is going to get lucky!

I jump up and down which is the signal for Andrew to get on the hotline to Mr No Bull. Mr No Bull, on call like an obstetrician, does not muck around. A few hours later he accelerates into our driveway with his vials, syringes and straws of semen-infused, manly little milkshakes, kept cold in liquid nitrogen.

Mr No Bull is a man, who, when asked at dinner parties what he does for a living, must choose between telling the truth: "hand jobs for other species", or lying. Either way, he's known as a manipulator.

He carries a catalogue with him of sperm donors. Pages and pages of cattle centrefolds, perfectly fine to leave on your kitchen table, no need to hide them under your bed, kids can even take them to school for show and tell. We can choose the breed from a huge array.

Black Angus? No, too angry hamburger.

Brahman? No, too lumpy, too humpy.

Charolais? No, too meaty, too whitey.

Hereford? No, too ginger, no ninja.

Highlander? No, too horny, potentially gory.

Miniature Galloway? Yes baby, yes!

And we can choose the sire's height, weight, eye colour, and even see his full name, giving us both the family tree of heritage and a glimpse into our own herd's future. Names like Castle Douglas Mickey, Wannawin Chocolate Soldier, Glenayr Cannon Boy. There's no anonymity here. No chance of accidentally sleeping with your own uncle or cousin or brother.

But what do we want in a father?

Like my own father, I want a magnanimous giant stuffed into a short frame...but without the bad jokes.

Like Andrew, I want someone handsome and broad shouldered, who smiles with mouth and eyes, someone who tightropes gentle and strong.

Like Jack, I want someone searingly honest, youthfully tender, funny and loyal.

And our herd might like to gene splice with a bit of bovine Jackman, Springsteen, Obama, Dalai, Channing, Clooney...with the concentrated DNA extract and talent of Tim Minchin and Waleed Aly.

And there he is!

On page 14!

Castle Douglas Mickey.

Eye candy for cows! Brain candy for bovines.

He sports a Liam Hemsworth, ready for filming chest, and peering out beneath the tumbling curls on his forehead are the sweetest of eyes delivering the most thoughtful gaze on four legs I've ever received. What else does a girl need? Whatever it is, he's got it: solidity in stance, flexibility in the flanks, an aura of calm...and the shortest of legs.

One Castle Douglas Mickey manly-midget-of-a-milkshake for our girl please! Extra cream!

And then we settle in for the 274 day wait.

But it's only 256 days.

Bellow!

Running, running through the long grass of the maternity paddock. Rushing, rushing to witness furry, knee-high new life. Camera ready to capture cuteness. Camera banging, banging at my side. Slowing, slowing so as not to disturb mum. Puffing, panting.

Sickening feeling rising, rising.

Bellowing. Bellowing. Turning, turning. A baby in full white sac, on the ground. Stilling, stilling.

The calf is trapped, forever in time. My heart is too. But my body doesn't hesitate, fingers tearing at the wet, waxy balloon of enclosure. I pound and rub and rock sodden hair. But there is no life left here, just damp, cold, slippery none-ness.

Just bellowing.

I have bellowed like that. The last just shy of twenty weeks. So perfect on the ultrasound. "Except," said the Doctor, "for the brain. Your baby may survive a few days after birth, but that will be it." I clutch Andrew's hand as the doctor continues, gentle of voice but ruthless, no rhetoric. "It's a long time to wait...for that result. You could...you could actually make that decision now."

But Andrew and I couldn't. Or wouldn't. Nature decided early for us anyway.

No heartbeat.

Bellow.

Life all around. Sun shooing off the last of the dew, gentle breeze quivering dandelions, willy wagtails hopping and preening, crickets catapulting. Camera lying in the grass, lens cap on, no memories to capture here. Sticky, slicked hands. A cow in mourning, transporting me back to mine.

Tears for a cow. Tears for our other four miscarriages. Intense gratefulness for the three who finally made it.

"It won't always feel like this," I speak across the species, soothing, sadding, willing, begging. Begging the promise of life – back to life.

Bellow.

"It's not your fault."

Bellow.

"We'll bury him and plant something special in his honour. You won't forget. I won't forget."

Bellow.

It is too much to bear.

Her grief.

My grief.

The Earth's grief.

Humanity's combined wails of loss, combined with the sorrowful farmyard bellows, clucks and bleats of the 56 billion farmed animals, once-were-babies, disappeared down throats by humans and our pets each year.

It's amazing our planet doesn't fall apart with grief. Fall into full eclipse. Fall out of orbit.

Maybe that's what storms, earthquakes and tsunamis are, planetary outpourings of sadness, aneurisms of anguish, the uncontrolled fracturing of a giant heart, the spilling of uncontrollable tears.

But hearts want to heal, rainbows want to arch, the earth wants to spin – untethered of the aching axis of loss.

Three living miracles have helped suture – though not disappear – five wounds for me. "It just takes time," I tell her.

Like a neighbour, I bring her food, biscuits of lucerne hay. And a steady stream of child mourners carry buckets of oats, tubs of bran, handfuls of molasses-soaked barley. We know it won't ease her pain, but at least their visits distract her with company and digestion.

Andrew collects the baby. Carries him in his arms. I try and collect myself, carrying myself home where I bury my head in 480 pages of 'Food Plants of the World', 480 pages so I won't have to witness the procession, the grave digging.

I've tried to dig out of grief before. But it was impossible to shovel. The dirt so heavy, so cold, so thick. A collapsed mine of darkness. Searing underground suffocation. Yet... imperceptible to human eye and heart, like dirt exposed to water, to air, to sky, the wracking weighted grief slowly begins to weather, to erode. A wind-carried speck here, a finger pinch there. Never enough for a handful, or a bucketful, and certainly never ever enough to fill a wheelbarrow. But, over time, enough that my surface was able to resurface, caked, but no longer buried.

I flip pages, reaching for the right tree to dig in above him.

Apple. Apricot. Avocado...

His life, to now provide for life.

Banana. Black Olive. Brazil Nut...

The circle of life, encircled by grief and gratitude. Renewal. Orbits as old as creation.

Page 126. Carob...

Carob. *Ceratonia siliqua*. Native to the Mediterranean. The seeds in ancient times used as a measure of the purity of gold, one "carat" equivalent in weight to one carob seed. Drought tolerant. Hardy in a variety of soils. A legume able to add nitrogen to the soil, thus improving it for other species. Evergreen. Shade providing in the heat of summer. A large, annual harvest of an energy-rich, chocolatey pod, able to be eaten by humans, cattle, sheep, goats and donkeys. Life saving. Long-lived, to 200 years. And I've read somewhere else: the tree you plant so your grandchildren, and their grandchildren, will never go hungry.

The tree you plant to continue life.

I want to order it. Who wouldn't?

It's a he-she. A grafted, self-fertile variety. A carob hermaphrodite called Clifford. A non-gender specific Carob who will grow to seven metres.

Click. I do.

Outside, I see Andrew's sweaty shirt-back returning the five-foot-tall, clay breaking crow bar to the shed. He slumps on the death shovel, pulls out his mobile. I bring him a stainless-steel cup of cold water.

"Next month? Sounds good."

"Who you-?"

"That was Jim, he owns Castle Douglas Mickey."

Long.

Controlled.

Sip.

"Mickey's coming to live with us," he says, "...so he can actually be with the girls."

Like me, my husband's buying in hope. My healing's in the shape of a carob. His is in the shape of a take-control-of-this-situation bull. And the kids' healing is in the shapes of the bright white filaments of dandelion they're blowing into the sky, and the little toy car and old teddy bear ornament they've buried with the babe.

A month later, our new babe exits his vehicle, hope re-entering our lives. Castle Douglas Mickey, the pin up boy, the ultimate brainy brawny beefcake from Page 14 of the sperm catalogue, is here, ready for his grand entrance. Ready to entrance.

He stands solemnly at the top of the ramp, dressed for the occasion in regal white fur coat, bathing in the shampoo afterglow of a five-litre bottle of Sullivan's Kleen Sheen for Cattle. If the girls were like angels when they arrived, he is like Adonis. Backlit by the setting sun, he practically shimmers.

Such a gaze.

Such a chest.

So much better in the flesh.

Andrew, Jack and I, the welcome committee, cannot help but be awed.

Castle Douglas Mickey has such style. Castle Douglas Mickey has such swagger. Castle Douglas Mickey has...oh dear...begun to pelt down the ramp. Who would have thought those stumpy legs could run so fast?!

This is not the respectful, dignified, in memoriam entrance I was thinking of.

Castle Douglas Mickey has started to run. To let loose. To give chase. To let it all hang out. There is no procession. There is no ceremony. There is no bowing of head or solemn reading of epitaph or reverential acknowledgement of his IVF wife's loss. There is just bull on a mission. Stalking, shuddering steak. Gonads on safari.

"What's he doing Mum?"

What to tell a five-year-old? Saying hi? Playing catch? Saluting?

But much easier to answer than "why don't I have brothers?"

It wasn't meant to be? You nearly did? You did?

"Well Jack," I say, um aah um aah. "It looks like he's…" Gulp. Swallow. "Having sex."

"Lots of sex," he observes.

Manly little milkshakes are being enthusiastically home delivered to customers right across the paddock.

Jack watches. Waits. Watches some more. Thinking.

"Did you jump up on Dad that way?"

Gulp. Swallow. This is sooo not respectful. This is an in your face, no getting out of this one, 4-D birds and bees chat, brought on by 600kg of frothing, licking, milkshake delivering, humping, grinding, lip-curling bull.

"Well Jack," I say, um aah um aah. "It was kind of the other way around."

He watches. Waits. Watches some more. Mickey's now doing another speedy home delivery, and I'm beginning to worry Dominoes will headhunt him for their delivery crew. I'm also worried Jack's going to ask me more about positions, aren't kids meant to use the internet for that?

"Mum, can we go now? This is getting boring."

"Sure," I say. Phew. And as we move off, forward, holding hands, tiny particles of grief loosen and lift, usurped by fine dustings of gratitude, sprinklings of renewal.

Before next summer I wish for green shoots on the carob.

Before next summer, I long for Fun Factory to be a mother again.

Before next summer I ache for of paddocks of cuteness, families of fur with fluttering love-me eyelashes.

I kneel, collect a dandelion blossom, gently blow. Filaments float skyward, my mind with them. Hope rising.

* * *

I HOPE YOU GET ENJOY THE CAROB/RAW CACAO RECIPE ON PAGE 251 with loved ones.

JAMMIN' WITH BIKIES

It's bat dark as Lucy and I head west from the farm. She's nine and even after a year the monthly 5 am drive to Gloucester Farmers Market still seems like a treat. For the family, it's both a joy and a necessity: a dollar made, a dollar repaid, that's why the snooze button never gets whacked and why we're now doing four different markets a month.

We glide around safety-fenced curves, slide over nubby, stubby hills. Headlights beam and bounce off the soft fog. We whizz past Dogtrap Creek. Today there are no stray cattle on the tar, no bonnet-bounding roos or brake-inducing foxes. Not even Pat the Jam Lady is ahead of us. That's a good thing, because lovely Pat has a puttery, ready-for-replacement, accelerator knee. She also has a puttery, navy blue, ready-for-replacement "God Loves You" stickered station wagon complete with gazebo and market tables strapped atop like tortoise shell sails.

She'll always pull over somewhere safe to let you pass, but no Pat in the way today means I can go a smooth ninety kilometres an hour all the way, an unimpeded rhythmic flow.

It's eight degrees Celsius and not quite 6 am when we arrive first at Billabong Park. First is good at a market. It means you can manoeuvre into your site without swearing or being sworn at, it means you can start your marquee raising and table-topping before the stallholder chat-fest, which means you should have approximately one minute to go to the toilet before the market officially opens. Otherwise, you officially have to hold for six hours.

Even though I take the gutter slowly, the ute's pregnant-with-produce belly grazes it. But it's lucky it's slowed me down; the head-lights pick up the misty shape of a body ahead. And another. Then another.

Brake!

What the-?

Two figures stride out of the fog. Buffed. Leathered. Booted.

I hope it's the Ulysses Club, those revving road hog retirees who petrol into the market for lattes and sourdough on their 'wild' weekend rides.

Harsh thunk on the window. Like when you hit a pigeon. There's a leather-clad knuckle to my right.

I squint. My eyes move to the patch on the leather jacket now air-bagging the window. Some kind of bird... Sparrow? Hawk? Eagle? Oh, and a swastika. Great. A Swastika and an eagle. In the park. In the dark.

I try and make out the words. Is it *rich*? *Ich*? No...

Is it *Reich*?

Oh dear.

Sounds pleasant. I'm guessing it's less a motorcade of mopeds and more a petrol-fuelled procession with Hitler at the helm.

Too late to reverse.

Bummer.

"What ewe think ewe doin'?" asks the short one who sounds like he's put a vacuum cleaner on reverse to throttle air up his chest.

"Um, ah, um, ah...there's a farmers' market on here today guys."

I glance at Lucy. The family's junior debating champion has nothing to say.

"Second Saturday of every month. EVERY month." I gabble.

"Stay here," Hoover orders. He and Jackboot confer, chopper-style. The sound of dull chainsaws.

I confer in my head too, a little higher pitched, more NutriBullet-NinjaPro: *They're not going to kill the honey lady, they're not going to kill the honey lady. Bad PR. Bad PR.*

I'm liking this line of thinking. I roll with it. They must know that if they kill the honey lady, the cops won't be able to turn a blind eye to their meth deals, the standover stuff. They kill the honey lady AND her cute daughter who taught herself to stand atop a racehorse while cracking a whip, and they are so, so stuffed. The media will eat them for breakfast. On toast! Thick bruschetta type toast! Whole loaf of week-old Russian black bread toast! They are NOT going to kill the honey lady. Or her daring, thoroughbred-taming, pre-tween daughter.

"Where's your spot?" demands Hoover.

My wary, wavery finger points to the 3m x 3m patch of grass between the garbage bin, the slightly slanting burgundy light pole and the walkway, known as the 'trip-way' by stallholders thanks to the not-to-be-denied tree roots pushing up the paving.

Right in the centre of my spot, a bikie, the size of a bull walrus, drools in a swag.

"You want *that* spot?" asks Hoover. "*You* go get him up then."

Hmmm, let me think about all the upsides of waking up a swastika-wearing bull walrus.

Thinking, thinking…

"I'm not waking him up," say I, the not-to-be-killed honey lady. "*You* wake him up."

Okay, maybe that was pushing my luck.

I now repeat rapidly in my head, like a NutriBullet-NinjaPro Rosary, "they're not going to kill the honey lady, they're not going to kill the honey lady." I squeeze and release the notches on the steering wheel like they're the little beads from St Kevin's Catholic Primary School where they kindly taught me to read, and as a reward, let me vacuum the convent at lunch time.

Hoover and Jackboot look at each other. I look at Lucy. She looks at me. More fog. Each time it rolls across the park I spot more swags. Empty bottles are piled up like garrisons. Fortified. There's been a naughty Nazi bikie sleepover.

"There are more farmers coming," I say. Breathe. Breathe. "And then," I pause for emphasis, "all the customers."

I can see them in my head now like they're the ANZACS coming to liberate me – the bowlegged farmers in their dusty and dusted up Akubra's, the tree-changers in their polished R.M.Williams boots and bright-checked shirts, the barefooted backblockers, the caravan park community, the Barrington Tops World Heritage area explorers, the genteel bookstore owner. And my favourite of all, the glowingly kind hearted ninety-year-old ex-beekeeper who sits with me each month, yarning good-naturedly until his oxygen tank runs low and the wheeze from the brick-dust factory of his youth wins out.

But the sun hasn't yet risen and all but the dairy farmers will still be tucked up in bed.

That'd be nice.

Instead, I am wide awake witnessing hung-over, pissed off bikies coffin-raising out of their swags. It's a bit like watching a mass zombie awakening: you don't know if they're going to eat each other, you, or come for your child.

It's the market of the living dead. A potential agricultural, artisanal, apocalypse. But bullet holes of sunlight and ute headlights begin to puncture the dark. Dawn has her ammunition.

Dave the Macadamia Man, Mel the Seedling Lady, David the Cheese Man, Lou the Sourdough Lady, Trish and Evan the grinning garden gurus appear ghost-like in the mist followed by the clinking and clanking of extending marquee poles.

Canvas swooshes over metal skeletons. Thud. Thud. Thud. Tent pegs drive into the hard ground. Tables unfold, tablecloths unfurl, breakfast sausages start to sizzle. We farmers unpack, the bikies repack. But they're not quite ready to be sent packing.

Hoover and Jackboot lurch back over. They lean heavily on my rickety, hardware store table – like they want the table to know they are there. They motor their not-so kid-gloved hands over my handmade organic lavender goat milk soap, the hand-blended honey vanilla lip balm, the hand-picked mulberries – like they want ME to know they are there.

Ah, it's kind of hard not to miss you guys.

Eventually Hoover strangles a 500 gm jar of tea-tree honey.

"Ow much?" asks Hoover, who in daylight is beginning to look more like a leather-upholstered, battery-operated, handheld dustbuster.

"$8."

"I'll give you," he pauses. Smirks. "Three."

Breathe. Exhale. Breathe.

Not working.

Breathe. Exhale. Breathe.

Still not working.

Oh dear, out it comes.

"You'll give me $8." Blurt. "I did a lot of work for this." Blurt. "This is hand-spun!" Blurt. "I'm nice to the bees." Blurt. "It's not mixed with imported gunk like the supermarket stuff." Blurt. "Or corn syrup." Blurt. "We're a small family farm." Blurt. "It's delicious." Blurt. "It's raw."

In my head, I don't hear or see a patched bikie, I just feel the spine-snapping weight of the hives, the hairdryer heat of the honey shed, the days spent cranking and bottling and labelling. The bills for the jars. The bald tyres on the ute. The invoice from the vet. My little bees hauling the pollen and nectar like underpaid flying couriers. Me, hauling it around again. For less. The smell of rotting straw mixed with sweat. The invoice from the insurance company. Armpit stains. Hunched, aching back. Hungry, stamping horses. I...I feel like smiting Hoover and Jackboot down with my gumboots. Feeling brave? No! Feeling mighty self-righteous? Oh yeh! Call me Minister Mellifera, aka Honey Bee, from the Church of Sweetness and Get Stuffed.

They watch all the backstory in my eyes, IMAX-style.

Jackboot hands over eight.

Hoover looks unplugged. Cord retracting.

7.47am. The bikies rev and roar so neither Dave the Macadamia's man railing against Monsanto, nor Lyz the flower lady's juicy morning burps can be heard. The smell of burning oil pancakes Billabong Park, the fog returns as fumes, there's a final Harley crescendo, a death metal, rolling, ribbity sound like giant cane toads being gnashed in engines. And then…the tension guns away on fat black tyres.

I plonk down on my fold up hardware store chair. P-lonk.

A little old lady in a crocheted cardigan, pink roses embroidered across it in the sweetest way, gingerly lifts an $8 jar.

"Your stall looks lovely darling, how much?" she asks sweetly.

Now I feel bad. Like she'll think $8 is overpriced, even though I know it still won't pay the bills.

"$5 for you." I say.

"Thank you dear," she says. As I put the $5 in the tin, she puts the $8 jar down and deftly slides the much bigger, higher, wider, heavier, mighty-hard-to-miss, actually-it's-nearly-as-big-as-a-wombat $16 one into her pale pink, reinforced shopping bag.

Nothing comes out of my mouth. I look away. She totters away. I look back. Pretend to myself and to her I didn't see. Don't see. My cheeks pink like her roses. I look down.

On myself.

Dissing bikies but diddled by the wolf in grandma's clothing. Some how I've let her steal our honey. And my voice.

My brain swirls with a life-time fog of middle-class, be kind to others, good girl programming…

Give everyone the benefit of the doubt: she must have poor eyesight, maybe even dementia.

Help thy neighbour: she couldn't afford it and needed it for her bed-ridden, pneumonia-fevered friend up the street.

Respect your elders: ahhh…is it okay to even slightly, minutely, tiniest speck of lint think a granny might be a self-serving, entitled, thieving bitch?

Her getaway is at pensioner pace. The chase would be over in, well, Olympic record time. But I remain rooted behind my laden tables. Laden by the theft of effort, care, hope. By crumbling romantic notions of the trust between earth, farmer, end-receiver. The trust between unborn, living, nearly dead. The trust in myself to get the job done. The trust in...okay, I've never trusted bikies.

The local storekeeper Malcolm once offered me whispered advice, "watch out for the old ladies," he'd warned, "they're the worst."

I'd laughed him off, thought him jaded. No way! The stalwarts of society: shoplifting seniors? Jerk geriatrics? Gangster grannies? It's the bingers, the bashers, the whole body-tattooed you need to look out for, not the wrinkly-bosomed, Johnson's baby-powdered, been-a-diabetic-for-thirteen-years-now Beths, Barbaras and Bettys...a group of which I too will be a member of in the not-too distant future... minus the controversial baby powder of course.

I keep looking at the space where the jar had sat as solidly as my attitude. I focus on the invisibility of loss. The potency of vacancy. The vapour of past, present, future. $16 of air. $16 of me being too embarrassed to stand up for myself, for my family.

"Mum, I'm cold, can I buy a hot chocolate?" asks Lucy.

I clink together enough 20 cent pieces to warm her up.

"Stall fee time," smiles the market organiser, invoice at the ready.

I fumble together a fistful of $2 coins to earn the receipt.

"Would you ever so mind donating some honey to our raffle for the Bowling Club?" asks another sweet old lady accompanied by her pink-tipped walking stick. A stick she doesn't seem to lean on so much as to waft and wave like a wand.

I'm in her spell. More childhood programming means I can't form the word 'no', so I say "of course" and clank two jars of Coastal Grey Ironbark across the table into her expectant bag. My Yin feels generous. My Yang feels 100% patsy.

Swindled by the wrinkled. Emptied by the entitled. Muted by me.

I can feel it; my big wide forehead is frowning. I adjust my apron, I adjust the honey, but I can't adjust the frown. Why should I pay for them to go bowling? Do they even know how much farmland and housing costs these days? That you need two and a half salaries to cover a mortgage, not the one that worked in their day? And why don't you donate yourself seeing you just told me you won't be at the market next month because you're heading off on a twenty-one-day fuel-belching, stomach-extending cruise? Give me a break! Can't they see anything beyond themselves? Like, how the planet is stressing? Like how I'm stressing?

My Mum, born in the thick of the Great Depression of 1929, caught by Polio in 1935 and rationed in WWII, spoke with her supple mind from her crippled body when she said, "darling, I think our generation has had the best of it. Used the best of it. Sorry about that."

I want to save the Red Riding Hoods, those I know and those I don't, but the wolf in Grandma's clothing is everywhere. I see it in my own wants, in the products we choose to buy, the fuels we use, in boardrooms across the western world, in governments, in a pink cardigan with roses.

On the ride home, our measly takings rattle in the tin at sleeping Lucy's feet. My head rattles along with it. How will Lucy ever afford her own home? Rattle. How high will she have to climb to avoid rising seas? Rattle. What whips will she have to crack? Rattle. What whips will be cracked on her? Rattle. Will she have to get up this early forever? Rattle? Face bikies, and grannies, alone? Rattle. And yikes! What kind of granny will I end up being?

Rattled.

I'm in love with the idea of the sustainable legacy we want to leave the kids, but today it just feels like a leg iron. Our ute, belly still full of the produce we didn't sell, crawls along with my thoughts. Rattle. Behind a convoy of cars. Rattle. Behind a cattle truck. Rattle. Behind a conga of seniors on Suzuki's. Rattle. Behind Pat the Jam Lady.

Traffic jam on the road.

Traffic jam in the cashflow.

Traffic jam in my head.

* * *

MULBERRY JAM ANYONE? THERE'S A RECIPE ON PAGE 251.

OPEN THE GATES

*T*he more time we spend with guests and at markets, the less time we have for our actual host – the farm – and it's eating me alive. Though bucolically beautiful to look at, it's wretched within. Centuries of heavy coastal rains and farmer-lit fires have leached nutrients from the soil like a reverse Berocca. Razor grasses grow so sharp they grind the old sheep's teeth to gums, and rotting fence posts and sagging wires are the only invitation the animals need to inbreed. It's paradise lost, and I desperately want to find it again.

That's one of the reasons I'm waiting on the slatted timber seat at Taree railway station for our first Wwoofer since the Germans; it's an admission on two fronts: we can't do this alone, and it's not like we're earning enough to hire someone. Oh yeh, and I'm a sucker for a sob story. We're putting our toes back in, but hoping not to lose our legs.

"The XPT service from Sydney is expected to arrive in ten minutes," says the tinny female voice over the speakers, and I cross my fingers in the hope the exchange will turn out well.

Her Canadian voice had been wavery, on edge, "I found your farm-stay on the internet," Robin had said on the phone a few days before, "I'm in Bondi and need to get out of the city. I want to be on a farm with a family for a while, just ten days. Can you help?"

"We're not a registered Wwoof host," I explained, getting ready to decline.

"Please, I just want food and board and I'm happy to do whatever. I love animals and I'm a graphic designer too, I really need to get out in the country. Please."

The Mum in me went maternal at the pleading sound in her voice, the business person in me went *wow, we need a new logo*, and the farmer in me thought *we might actually get something done*.

It took two days to get the family on board, Rosie the last to relent, and only when I'd promised her for the hundredth time it was definitely not the second coming of the 'Dermans'.

Now here I am, watching the sleek train glide in. The doors open with a gentle hiss, and a smattering of old ladies disembark, wheeling floral and pale peach paisley suitcases. I wonder briefly if there's an ill-gotten jar of honey in any of them. Shoeless smokers jump off, light up and drag nicotine quick and deep, furtively checking for the stationmaster in the hope they'll get their fix before having to board again for the journey to Queensland. But where is Robin?

That must be her! Last carriage, pink-haired, backpacked, tentative like she's wondering what she's got herself into. I've been in her shoes, arriving somewhere new, so head straight up and give her a welcome hug. It's funny how you can instantly trust the goodness in someone with just a hug and eye contact, while with others any good vibes you hope for evaporate on contact.

"Thanks for taking me in," she says.

Over the next ten days she endears herself to kids, goats, Andrew and me with her gentle ways and tinkling laugh. I drive her crazy with my indecision over how the new logo should look, but eventually she nails it with a goat, hills, three starry bees and a hand-drawn circle of life. The logo sorted, we get busy making goat milk ice cream, goat milk soap and compost piles of goat manure that will go on the garden next season. We feed Robin up, the farm feeds her soul and she leaves brighter, happier and more confident than when she arrived. Ditto us.

"It worked! She was great!" I say to Andrew.

"It went pretty well," he agrees, "reckon you can get me a bloke to help with the fencing?"

And with those words the gates are open. I head home and sign us up to be official WWOOF and Helpx hosts and our ad goes up on their websites.

Growing up the youngest of six kids, I'm used to living with multiple personalities. As a teen, I spent a year as an exchange student, re-learning how to fit into three different Californian families. In my twenties, Andrew and I lived on a coral cay – half the size of the farm – surrounded by sharks and shoulder to shoulder with one hundred co-workers, three hundred guests and three hundred and fifty thousand flapping birds who lodged themselves in trees, burrows and your armpit if you weren't careful. When it rained, you could barely breathe for the people crammed inside and the gas from the ammonia-laden guano rising outside. I suppose it was all good preparation for what we are about to embark on.

A steady stream of short-stayers begin to arrive at the farm for five to ten day stints. It's like the United Nations convenes in our lounge room and we have to quickly navigate – without offence – personalities, dietary requirements, risks posed, skill potential, tool and animal-handling proficiency, wheelbarrow tenacity and potential gardening endurance. In between training and tour guiding them, we need to Dr Phil between the various nationalities, genders, relationship difficulties, unrequited loves, mental health issues, work ethic and expectations of Koreans, Japanese, Chinese, French, Italians, Fins, Swedes, Americans, Brits.

"The problem is," says Andrew, after a long day interpreting rather than irrigating, feud-solving rather than fencing, and giving 101's in tool use rather than wielding them, "it takes me a week to teach them anything, then they're gone."

"But if we take them for longer?"

"And we don't like them?"

"We're stuffed!" My shoulders slump thinking if the Derman torture had stretched to three months.

"But if they're good?"

"It would be amazing!" My mind runs away with images of the greening of the farm, the bars of soap and pots of balms we'll make, the towering sunflowers being harvested, golden honey being spun and bottled, the kids learning new languages and cultures. I think about the bigger picture: the opportunity for us to share practical knowledge of sustainability, organic agriculture and animal husbandry, enabling these tourists to incorporate what they learn in their lives for ever more...little ripples of eco-evangelism spreading out across the world.

Andrew's obviously been envisioning too because he says, "let's do it then, let's try a three-monther."

A week later Casper, a Swede, and Yan-ting, his Taiwanese boyfriend arrive. Twenty-something white collar workers in search of a second-year Australian VISA, they need to do three months' farm work to satisfy the government requirement. They'd picked our farm because they didn't want to go fruit-picking and wanted to stay near the coast. They've driven their own car and when they open the boot there are no backpacks, but a matching set of Luis Vuitton's.

"Hi, hi, hi," I say, a bit too enthusiastically, like when I met the in-laws for the first time – knowing we're going to be long hauling it and hoping we get along.

Casper and Yan-ting stand stiffly together so instead of offering a hug I shake their hands. Feeling the softness of theirs makes me realise how rough and sinewy mine have become.

"Do you supply sunscreen?" asks Casper, his pale complexion not even freckled.

"Sure, there's some on the table near the office."

"You supply soap?" asks Yan-ting, his face scrubbed clean and shiny.

"Plenty," I laugh, "you can help me make it."

"You supply juice?" asks Casper.

"Sorry, that's one of the things we don't," I say, having watched three, three litre bottles a day drained by the last two Wwoofers and deciding real oranges and plain water wouldn't bankrupt us, their teeth or cover the planet in plastic containers quite so quickly. "Junk food and alcohol are up to you too, but everything else is provided and you can let me know if there's a special meal or food you'd like."

"How many days we get off?" asks Yan-ting, his tone sharper since the juice drought was announced.

"Two," I say, thinking this was all covered in the phone calls and in the letter they'd already agreed to, "and you can choose if you want them on weekends or midweek, together or apart, we don't mind."

"We'll be taking them together," says Casper sternly, our welcome conversation fast going the way of a souring union negotiation.

"What that noise?" asks Yan-ting. "A monkey?"

"It's a kookaburra."

"Good, don't like monkey."

"Don't worry, the only monkeys you'll find in Australia are in zoos," I reassure him. "You okay with other animals?"

"Don't like animals."

My stomach starts to feel heavy, and so does my heart. We're only five minutes in, with two months and twenty-nine days to go. I try to come up with a joke...a Swede and a Taiwanese walk into a farm... but the punchline's on me so I say, "how about you unpack at the cottage, then meet me back here and I can take you for a walk around."

I busy myself in the kitchen, scraping previously spilt beeswax from benches and the floor. It's soft, pliable and comes up easily, kind of how I was hoping these Wwoofers would be. When I finally look up and out the window, I see two doctors, in full blue hospital scrubs, walking down the driveway.

Andrew could be anywhere on the farm so I ring him on the mobile.

"Babe, you expecting vets or doctors or something?"

"No, why?"

"There's just some people in medic or hazmat gear coming down the driveway. District vet maybe? You heard of any outbreaks?"

"No, want me to come back? I'm down at the dam fixing the solar pump, ants nest bunged it up. Ouch!"

"Nah, it's right," I say, my mind starting to churn through all the catastrophic outbreaks this visit might mean: Anthrax and Bird Flu and Bluetonge oh my!

Or maybe Screw Worm maggoted its way in from Papua New Guinea. Or someone used a private plane to smuggle in a dog frothing with Rabies. Or maybe the flying fox colony up north has moved south with deadly droppings of Lyssavirus. Geez, hope it's not that damned picornavirus, Foot and Mouth Disease hasn't reared its blistery head in Australia since 1872 and I'm not wanting to be ground zero. Hell! Not varroa mite, the poor bees!? Yikes! Things are looking grim.

I slip out of my non-slip, hose-able Crocs into my serious wellies and stride across the lawn to hear the fate of our beasties and the planet. "Hi, can I help you?"

They turn around to face me and the colour drains from my face.

The voice of an Emperor comes from the robes of a surgeon. "We are ready for walk," says Yan-ting.

I think something just popped my lungs. I think they've been popped because I suddenly can't find any air. Now I can feel my eyes popping, and now my brain is starting to pop. Like pop rocks. Pop-pop-pop. Pop-pop-pop. What the pop?

"What's with the medical gear?" I finally pop out.

"My uncle, he work for hospital in Sydney," says Yan-ting. "He help us stay clean."

I look at Casper to see if he thinks this is a weird idea, but I see he is totally on board, his scrubs laced up, each tie bowed like a professional, his baggy blue pants skimming his ankles. Snap! He flicks his surgical gloves and my eyes are drawn to the man-hair trapped beneath the latex. Pop-pop-pop. Anthrax, I was prepared for, but farm help in hospital scrubs? Any visitor will think we're running an insane asylum, or engaging in illegal body part harvesting, or so wealthy and with animals of such ill thrift we employ two vets full time.

My eyes are obviously still popping because Casper says, "and I want to protect my clothes."

Like that doesn't make it any weirder, I think, then scope my op-shop jeans…the worn knees, the barbed wire rip, the grass stains. Maybe they're onto something, maybe they're the sane ones.

Over the next few weeks Andrew and I grow ever more immune to the operating theatre garb in the paddocks, their clinical hold of their emotions, and thin scalpel lips that need to be stitched open to smile. But we can't get used to Yan-ting's vocals whenever he loses sight of Casper. His voice loudspeakers around the farm, no subtle paging system for him, no mobile phone, no going to find him or waiting for him to appear, just a blaring:

"Cas-PAARRR! Where you?"

"Cas-PAARRR! The chicken is near!"

"Cas-PAARRR! Get it away!"

"Cas-PAARRR! Come!"

"Cas-PAARRR! Cass-PAARRR!"

Every time I hear the first part of his name "Casp-" , my brain shouts "ARGH!"

It might be summer but it seems a frost has descended on the farm, even the kids read the human weather patterns, feel the chill.

"Why don't they pat the animals Mum?" asks Rosie.

"They're not really animal people darling."

She looks at me, wide-eyed, "How can you not be? Animals are so cute!"

"They only care about each other," says Lucy.

"They don't drive the quadbike very well," says Jack.

"I don't drive the quadbike very well," I laugh.

Later out of earshot of the children, Andrew expresses his concern too.

"Three times now they've left the two coop doors open, could have lost all the chooks to foxes. They've got no empathy," he pauses, "for the animals, kids, what we're trying to do."

I take off my rose-coloured glasses and put them on the bedside table.

"Yep, they're just here for the visa. They needed something, we needed something. But we committed to the full three months so we've got to make it work."

"I'll put them on non-living things," he says, and he does.

Months ago, we'd been inspired by the TV Show Jimmy's Farm. On it a friend of Jamie Oliver's was using a stationary caravan as a henhouse. After reading more, Andrew thought replacing the floor with steel mesh would let manure drop right through into the paddock, saving us the sweat of collecting it. And by keeping it on its tyres, we could tow it to anywhere on the farm needing fertility. The chickens would free range all day, then have night-time security from foxy loxy in their locked van. Driving down some backroads we'd spotted a bomb of a caravan languishing in a paddock. The farmer was happy to see one hundred dollars and the back of it, and it had since sat in our driveway waiting for renewal. Your time has come baby!

"You want us to turn it into a chicken house?" groans Casper.

"I not like chickens," whinges Yan-ting, not acknowledging his three-eggs-a-day habit relies on them.

"Well guys, I'm kind of stumped," says Andrew patiently, "you volunteered to come here, what would you prefer to do? What job would suit you?"

"Cas-PAARRR! Where the hammer?"

"Cas-PAARRR! Where the paint?"

"Cas-PAARRR! Where the rag?"

"Cas-PAARRR! Where you go?"

Andrew shows them how to use tools, stoically re-does their errors, teaches them to paint, teaches them to re-paint. Their never-ending supply of fresh scrubs come in handy as paint smocks, and the mouldy old caravan slowly rebirths in a wave of bright orange for (sunrise and hope), lime green (for earth and life) and blue (for water and sky) ...our colour choice, not theirs.

Their three months finally ends and relieved, I sign the Australian visa paperwork. They're not interested in seeing the chickens move in, or helping Andrew find any of the tools that have been lost, or saying goodbye to kids or animals. They're off to the city, back to computers and cleanliness and stuff that requires no beating heart. But they've helped us achieve something, they've taught us patience and there's a celebratory sense of relief as the echo of "Casp-argh!" finally fades.

We make the final touches to the caravan, Andrew replaces the broken windows with shutters and collects branches of Camphor Laurel for perches. It's a declared noxious weed in New South Wales but its timber is a brilliant repeller of mites and lice.

"I christen you, *Hens on Holidays*," I say, pretending to launch the caravan with champagne.

It squeaks and clanks as Andrew begins towing it to its first port of call, a paddock where the topsoil is so nutrient deficient it grows only blady grass – but hopefully not for long with the living fertilisers we're sending its way.

Thirty-six chickens perch like first class tourists, peering out windows, cluck-cluck-clucking as the view of the valley rolls past and they head toward the sunset. I feel my heart swelling, my smile broadening, it's both a gift and a grind to renew this land, but there's a little bit of paradise in every step we take…even if a few are backwards and with company more surgical than us.

* * *

A GREAT USE FOR EGGS IS TO MAKE FRENCH TOAST. THERE'S A RECIPE on page 252.

DOMESTIC UNREST

oday I just want to be invisible. I want to be the stick insect on rough bark, the pale gecko stuck to the cream of the eaves, the microscopic lime frog in the silverbeet. I want to hide from the frenetic ecosystem we've created of guests and customers and backpackers and animals and plants and bees.

Today I just want to be deaf. I don't want to hear the phone ring, the goats bleat, the wails of wilted leaves. I don't want to hear the grinding crank of the honey spinner, the cost of hay or how much sea ice has melted.

Today I don't want to feel. Not the heat blasting my face or a sting of a bee or a hoof on my toe. Not the pull on my hand to be somewhere I'm not, nor the ache of parental guilt or someone else's wants.

I just want to be invisible.

But Harry Potter took the damn cloak!

So out I go in my not-so-super hero outfit of dirty worn jeans, sweat-hiding black tee, boots and cap. It's 8 am and I'm in plain sight.

Nature spots me straight away. Nails me.

My ankle rolls in deep cracked clay. Baking westerly winds blow-torch greenery into parchment before my eyes. My eardrums hammer with the sound of animals pant-pant-panting in white hot shade. The

78

Rural Fire Service has shifted the arrow on the fire danger signboard to 'Extreme' and everyone is on tenterhooks awaiting the tell-tale signs of musty BBQ air and grey columns rising skyward.

The heat wicks not just perspiration, but joy. Oppressed by celcius. Exposed by degree.

It's been a tough slog, three days of blazing heat, two days of tricky Australian farmstay guests. There's one of them now, near the hay shed, it's one of the six-year-old twins. Uh oh, he's seen me. But worse, I see what he's about to do.

"Don't!" I yell. But he lights the match anyway, "don't!"

Done.

He propels the match from fingers to grass. The grass gulps flame like the water it craves. Ignition.

A glacier of melting lawn, rivulets of orange flowing low and quick. I stomp and slide, stomp and slide, scuffing flames into the earth.

I hear it again. Smell it. Potassium chlorate striking red phosphorous. Volatility in a tiny stick.

Grass being consumed.

Stomp. Slide. Stomp. Slide. If I don't stop him soon, the flames will get away from me. To left – to right – to the hay – to the neighbours – to the forest – to the 6 o'clock news. Geez, not the 6 o'clock news!

I erupt. Lunge. Roughly grasp the matches. The box of red-headed aspen timbers scatter across the ground. I stomp and slide on the last of the embers till there is just ash, till my twisted ankle burns.

"I will kill you!" he spits at me. "I will kill you!"

"I don't think so sunshine," I say, kneeling down to get eye to eye with him. "How about we play pick-up-sticks?"

Iceberg irises. Cold war.

Yep, he really wants to kill me. He is not in a pick-up-sticks kind of mood, he is in a let's burn down the entire farm and all the people and all the animals kind of mood. It's a staring, glaring, combustible male version of Carrie on her prom night. I wonder if he's related to Hannibal Jnr.

Twin #1 is doing my head and heart in.

You never quite know what you're getting when farmstay guests arrive. You hope for kind and interesting (the ones you can share vino and dreams with), you tolerate rude and boring (even if it's with teeth so gritted you need an emergency crew with the jaws of life to unlock them), but you take a full head and body slam when guests bring darkness in their luggage. This is why I want to be invisible, this is why I need a cape.

"Give them!" he demands.

Destruction is his goal so I know there's no point lecturing Twin #1 about how dangerous it is to light matches, so I simply say, "nope, I picked them up, I'm keeping them." And I shove the little Redhead box deep in my pocket. Deep as they can go. Actually, way down near my ankles and into the fat part of my big toe.

I wish someone could have shoved me way down deep yesterday when I had my awakening at the pool.

Kidney-shaped pool, clear blue tropical waters, creamy shade sails floating above, protecting. Palm trees and native frangipani dipping overhead; passionfruit vining near the fence, delicious baubles of summer. Chickens clucking contentedly, foraging near the gate. Relaxing sun lounges offering poolside escape from the mountain of farm heat. Oasis.

And the guests.

TwinMum: beautiful, raven-haired, diminuitive.

TwinDad: muscular, gregarious, Putinesque.

Twin #1: Seething. Seismic. Severe emotional dysregulation

Twin #2: Dress-wearing, dreamy, detached

"Why does he wear a towel hanging over his head?" Lucy asked me after noticing Twin #2's different dressing habits.

"I think he'd like to be a girl darling, the towel makes him feel like he has long hair."

"Oh," she thinks about that, no judging, just thinking.

I'm thinking now too…selfishly…how do I get past the pool and the guests and into our house unnoticed? I've been out filling water troughs and making sure all the animals have shade. I'm hot. I'm bothered. I want to chill out doing paperwork, not make polite conversation. Can I snap off a bush and slow step behind it? Should I army crawl? Do I use a horse for cover?

I mean, it's not like the dad's not a great conversationalist, it's not like the mum's not sweet, but they have let Twin #1 run wild and we've been on edge like the Rural Fire Service since they arrived.

"Come! Come in and join us!" says TwinDad, spreading his arms magnanimously.

Busted, gotta work on my army crawl.

Sometimes you just can't put your finger on what's wrong. Why do we feel guarded with some people? How do some people seem to dam our own ebb and flow? Why do we want to avoid rather than embrace? Is it cultural differences? Is it we're shy or they're overbearing, or vice versa? Or lacking in confidence or overloaded with ego? Or retreating or advancing? Is it just we vote differently? Earn, think and see differently? Is it the vibe? Am I just tired? What is it?!?

And then the answer, in this case, arises before me.

TwinDad lifts Twin #1 above his head. It's an edge of the pool game. It's fun. He's going to toss him into the pool and everyone will laugh. Just like my Dad did with me hundreds of times at the rockpool at Curl Curl beach.

But that's not what happens.

Twin #2 breaks cover. Screams! "No Daddy, no!" then uses his towel hair to cover his face.

TwinMum tenses, not lounges.

This is not fun in the sun. Though the dad jests and pretends it is. Showman. Shows me. Twins in terror. A reaction to action. Domestic violence. Dark control. That's what he brought in their luggage. That's what he's throwing around in the pool.

The plug has been pulled on the oasis.

I want to army crawl out of my skin.

What to do? In that instant?

Maybe I could stand up, fully clothed and wha...wha...whoah overbalance into the water. Farmyard jester. Lighten up the luggage. Dilute the domestic. Disappear into the chlorine with flailing arms and legs. Emerge to giggles. A temporary distractive bandaid when what's needed is full surgical separation.

Maybe I could stand up and say to TwinDad, "hey mate, looks like that's not so much fun for them."

Maybe I could drop down low, try a better army crawl.

But I'm melted to the chair, with liquefied lips and molten thoughts.

Splash.

Tears.

TwinMum towelling dry water, and salt. Times two.

"What problem?" demands TwinDad, "what problem?" He turns to me, louder again. "This is fun! Yes?"

"It can be," I say softly.

That night in bed, fan click-click-clicking overhead, Andrew says to me: "don't beat yourself up, we've only just met them, you can't know what's going on."

Yes, I don't know their story, but it's obvious it wouldn't be an easy read: ultra harsh upbringing, the pressure of two babies at once, unconscionable rages, unconscionable violence. Generational casualties.

"We should probably have done a course in psychology before we opened this place," I say.

"We're a dinky little farmstay, not a clinic," he replies.

I put a pillow over my head, the same head it seems I rented out when we rented the cottage. I wonder if the receptionist at the Ritz Carlton feels like me? Or the concierge at London's Corinthia? Would the butlers at the Burj Al Arab be spending the night fretting about the mental tonic they might have served? Would the laundry personnel at The Peninsula be perspiring about how to better iron out an ugly human crease? Would they? Or is it just me?

I shove another pillow on top of my head, the first one's not working.

"Let's just give them a good holiday," says Andrew, tapping on the pillow defences. "Help them relax." Tap. Tap.

Take the heat out of things.

Like I'm trying to do this morning by putting out Twin #1's inferno.

Done. Matches secured.

Now to try and crack the ice.

"How about you go grab your brother and Andrew will take you down to the tyre swing," I say.

No hint of a crack, not even a hair-sized one. Twin #1 could reverse global warming.

"The pedalboats are down there too, you could go for a ride."

Still no crack, no melt. Got to get him a meeting with Al Gore.

I play my hottest trump card, the child manipulation one, the tried and true that Andrew uses too. "Farmer Andrew might even let you steer the quadbike."

I hear it! A slight creak, a fissure, a glacial c-r-a-c-k. The offer of power and control has been accepted, his little legs gush down the driveway.

TwinDad is nowhere to be seen as the boys position themselves on the quad with Andrew.

"Want to walk down with me?" I ask the Mum, partly so Andrew doesn't have to wrangle the twins alone and partly so I can share with her the peace of the creek. "It'll probably be too hot later."

She nods and we begin to walk. "The kids are having a good time thank you," she says. "Your husband is very... patient."

"That's because he's had years of experience putting up with me," I say, knowing again in that moment the easy confidence I've been gifted by being raised by and married to gentle men. It's such a morass, the man woman thing, I feel guilty for my luck because I'm walking beside someone who doesn't share it, and at the same time I know it shouldn't have anything to do with luck, gentle should just be the way it is. Not could be. Or should be. Just be.

The oppressive heat and my brain temperature subside as we walk beneath the trees. Andrew has the twins on the tyre swing, pushing them only as high as their laughter, not further. When it's just him and Jack, he'll propel him twenty metres into the air, an almighty high swish, the grin on Jack's face captured by the wind, my heart captured in my lungs. But he knows these boys don't sail like Jack. Different waters, different winds. In need of safer heights.

I guide the Mum around the five-wire fence of the tyre swing paddock, around the side of the lily-patterned dam, through the tall crunchy Eucalypts to the creek bank. The heat subsides even more, the moderating effect of shade and water and reeds. The gentle trickle, trickle, trickle over mossy, long-fallen logs. Centuries of depth and flow.

"It has been good to be here," she says, looking at the moving water, not me. "We have had difficult times."

Trickle. Trickle. It all comes out.

I take her hand in mine. The current flows.

When they check out the next day, I go to the cabin and strip the sheets, plump the cushions, empty the bins, and finally, turn off the fridge to save power till the next booking. Atop the fridge is the guest book where people leave us comments, I always save looking at it till last.

"The most relaxing holiday we've ever had. Thank you from my heart."

Twelve words. Twelve words that make me feel so selfish. Twelve words that spell out my desire to be invisible meant I wouldn't have seen; my desire to be deaf meant I wouldn't have heard; my desire to not feel meant neither of us would have been touched.

Twelve words that make me feel guilty.

For still not wanting that. For not knowing how to handle all that. For not knowing how to make it all better.

I walk down the ramp carrying the dirty laundry. My ankle rolls again.

Boiling stress condenses.

Trickles.

* * *

FOR MINOR SPRAINS AND BRUISES, GENTLY APPLY THE BALM RECIPES ON page 221-222.

84

THE BEE MOTEL

I turn off the headlights, look skyward. The stars are putting on a show again and the Milky Way glows like the diamond sheen on a galloping black thoroughbred. I've just driven back to the farm after a meeting of the local tourism board, the news: the government's launching a one-off grant program for regional tourism operators. Though it's a longshot – about as long as Mars is from my eyeballs right now – if we come up with a great idea and it's selected, they'll match development funding with us fifty/fifty. The potential lifeline couldn't have come at a better time.

We eke out a living from the farmstay. We eke out a living from our hand-spun honey, our hand-milked goats, our handmade balms, our hand-poured candles, our hand-dug turmeric and our hand-spun scarves from hand-fed, hand-shorn, hand-shaking alpacas. We eke out a living on the crumbling crust of the breadline.

Eek!

Emotionally it makes me flip flop between feeling worthy and feeling like a sucker. Between thinking what we do environmentally matters and that nothing we do does or will. I flip flop between passion and angst and flail myself with three phrases: *We're not doing enough, we're not earning enough, we're not enough.* I'm not evolved enough to be happy with our choice of the simple life if financially it means we might lose it.

I fill Andrew in on the tourism meeting, finishing with, "maybe this is a sign we either give it a crack and get bigger, or we give up and get real jobs."

"Another farmstay cottage?"

I shake my head. "They said it had to be something innovative, original." We throw ideas back and forth.

"A corn maze!" he says. "They're huge in the States."

"They don't have as many killer snakes. Plus, how do we earn an income the other forty weeks of the year when it's not growing?"

"A tree house!" he says. "I've always wanted to build a treehouse." I don't need to say anything because his face drops as he realises the farm's clearing by previous owners hasn't left a single ancient tree to build in.

"How about a beetrootery and a meadery?" I say, "weird fruit and vegetable wines!"

"Drunks, bulls and barb wire...a cocktail for...?"

We go backwards and forwards till midnight, till we yawn louder than we talk. Around 1 am our verbal brainstorm snoozes, and so do our thoughts of the grant.

I'm still yawning at 8 am when the yellow phone starts buzzing to the *Ol' Macdonald Had a Farm* ringtone.

"Good morning, Anna speaking."

"Hey there, we're just up the road, can you do us a favour? There's a bunch-a-bees in our mailbox and the postie's not mighty happy. Reckon you can get 'em out for us before they get sprayed?"

It's not something I like doing, catching swarms. I've caught two in the past with Rod and though it's easy if they're low on a tree or on a post, the thought of bringing unknown diseases back into our apiary puts me off most call outs now. But it's a local, and it's a letterbox, and they'll be poisoned if I don't. Can't have that.

I suit up and drive up the road, it's an easy job, a matter of scooping the bees into my empty box, making sure I have the queen so any worker bees out on the job will sniff her out on return and move into the box too. I pop in to let the neighbours know I'll be back tonight to pick them up once everyone's safely home. Walking through their house garden I'm struck by the riot of colour and scent. From ground level to head height and beyond, flowers and vines in hues of pink, yellow, lime, orange, blue and amethyst layer and stack around me like a library of colourful botany. My nose goes blood-hound, sniffing bouquets of jasmine, mandarin, cinnamon, gardenias, rose, lilac, and another scent that evokes memories of my mother's perfume, the one in the precious crystal she only opened on birthdays and special occasions.

The scents flick a switch and suddenly all my senses are on alert. My inner ears trampoline to different sounds and my eyes zoom into focus. The revelations begin. I see things I've never seen before: voluptuous Kardashian-bottomed bees, iridescent blue chequerboard-patterned bees, helicopter bees shiny like fresh cut opal. Bees so tiny they could be midges on the wing, and bees so big they could be blowflies. Among them fly the familiar black and gold of the Italian honeybee, but how in all my years of life have I never seen these others? Where did they come from or has Nature decided to prop open and laser polish my eyes?

Amid the colour, sound and smell – the opposite of our bland, grassed paddocks – I feel like I'm being LSD'd by nature. I rush home and Google "Native Bees". There aren't many results but the ones I hit on amaze me, amaze me until 2 am, 3 am, 4 am. I'm still awake when Andrew wakes the kids for breakfast. I hit them with all I've got.

"Did you know there's two thousand species of native bee? That honey bees have only been in Australia since the eighteen hundreds? How come they never taught that in school? It was native bees who pollinated the macadamias and everything! And guess what…?"

"What?" Lucy asks, knowing I'm not going to be put back in my cereal box just yet.

"Some of the species are stingless! And guess what…?""

"What?" asks Lucy, she's on automatic now.

"If the honey bees die, we'll need native bees even more for pollination."

I stuff them with information for another half hour and it's the first time in weeks they've been happy for the school bus to arrive.

The discovery of the native bees planted a seed in my mind. Over the next week I water, fertilise and shine the sun on it until it shoots. I dream of it at night, and spark from it in the day. A week later it's taken root and vined its way into my every waking moment.

"I've got the tourism grant idea," I say to Andrew and begin to put the vision into words, first for him, then into forty-two pages of text, columns and graphs for the bureaucrats.

IMAGINE...IMAGINE A VIBRANT, EVER-EVOLVING FIVE-ACRE GARDEN FULL OF plants of importance to humanity. Imagine a beautifully functional jungle where all plants have a purpose: edibles, medicines, dyes, building materials, beverages, perfumes, insect repelling, fuels, poisons, fibres, essential oils, animal fodder, bee fodder, habitat. These plants are needed now. These plants will be needed in one hundred years and forever. And people will need to know of them.

Imagine set within this garden a mini five-storey Bee Motel set up for fifty different species of solitary native bee, each suite providing a unique habitat for locally indigenous yet mostly unknown bees such as Teddy Bear, Blue-Banded, Reed, Resin and Cuckoo bees. Imagine sprinkled throughout the sanctuary, colourfully painted hives housing families of the tiny ant-sized, social, stingless Australian bee, Tetragonula carbonaria, who produce the world's most delicious, medicinal and rare honey...but only two small jars per year. Imagine how experiencing this sanctuary will emotionally move people, providing them with an "a-ha!" moment as they understand and enjoy the connectedness of humanity, insects, plants, the earth and each other.

This project will not be museum-like, we dare to be different. Visitors will immerse themselves as they wander the gardens, seeing, smelling, touching, tasting, learning. They'll learn plants can be grown in the shape of walking sticks, plants can be used for toilet paper, plants can burn off warts; that there are plants whose scent will make you happy, whose nutrients will make you well, whose timber you can craft into flutes. They'll learn about the

plight of the honey bee, and how by providing habitat for native bees we can
improve the resilience of all. For all.

This small family-run attraction, set on a ninety-acre farm already
operating an award-winning farmstay, will be energetic, vibrant, ethical,
beautiful, nurturing, nourishing, thought-provoking, authentic, life-
affirming...

"Do we go for it?" I ask Andrew.

"One problem," he points to the financial section of the document.
"It says we need proof of matching funds. If you're including a shipping container for a shop, a deck, the fencing, all the plant signage, the
earthworks, the plants and labour, the native bee houses, the website
etc, the whole thing will be eighty thousand, you need to show we can
provide forty of it."

"Can we get a loan?"

"On our current income?" he shakes his head. "They'll see we can't
service it."

"But with the extra income the sanctuary would bring in
we could."

"They don't factor pie-in-the sky in."

"Can we sell something?"

"The farm."

"Ha ha."

I droop in the chair, whoever said money can't buy happiness
never wanted to do something this bad. Andrew stays silent, but his
fingers tap numbers into the calculator.

"We might be able to get another credit card. 18% interest but it's
probably the only way."

"Can I tick the *yes we can match it* box?"

"Might as well then, go on, press send. Press it."

I close my eyes, depress the key, slowly, purposefully.

A finger pushing hope.

The proposal for a tourism grant for a native bee sanctuary and
useful plant garden is away. Pages and pages of budgets, plans and

research, now email their way invisibly to the nation's air-conditioned capital.

Through the grapevine, we hear there have been hundreds and hundreds of applications from across Australia, from mega tourism companies right down to micro-minnows like ourselves. Hope fades quickly, even though my dream doesn't. I start concocting ways to raise the money to make it happen anyway...put in acres of saffron, rent the alpacas out as page boys for weddings, publish a calendar of our chooks in compromising positions, track down a billionaire bene-factor who's big into bugs, or a long lost, probably doesn't even exist rich relative who wants to make up for unfulfilled life in his last moments of it. Or ask a different bank, and then another and another, for more personal credit cards.

Life goes on at the farm, leaves from the liquid amber trees float to the ground and the first thin layers of winter frost coat them with an icy shine. The cold weather's a perfect excuse for inside jobs like soap and balm-making, and every spare minute I leaf through books researching plants, herbal medicines and native bees.

The fourth Tuesday in July 2011 starts as any other. Andrew brings in still steaming goat milk for filtering through cheesecloth and I pour it into glass bottles for our consumption, and into ice cube trays for use later in soap. I launder guest sheets, place a classified ad in Sydney's Child Magazine to attract more farmstay customers, then drive into Nabiac to collect the mail. Fancy that, a vet bill. Oh, and a hay bill. That's cute, a hand-drawn thank you from a five-year-old guest. There's a farm catalogue with everything from sheep halters to electric killing booths for rodents, foot rot shears to woolly coats for goats. And then I see the letter postmarked Canberra, ACT.

Everything's riding on this letter. All my fundraising ideas failed: we couldn't afford the saffron seed set up, no one wanted alpaca page boys, the chooks went all coy, and I never tracked down a bug-loving billionaire or wealthy relo even thirtieth removed.

I carry the envelope like a premature lamb to the car, put it on the passenger seat and seatbelt it in. On the way home, I drive slowly around potholes, give the roadkill roo a wide berth instead of picking it up for the compost heap, and steal sideways glances at the envelope to check it's okay. It is, it hasn't moved, but I can feel my blood on the move, surging unchecked through the veins in my neck. As I unlatch the gate my fingers fumble and flutter.

I track Andrew down in the animal feed shed, his sleeves dusty from mixing dolomite, seaweed meal, oats and bran.

"The grant announcement's here," I say, surprising myself by tearing up.

"And?"

I pass it to him to open.

His green eyes give nothing away.

"Thank you for blah-blah-blah. Many applicants blah-blah-blah."

I crinkle my toes in hope. I think of the knowledge we could share, the ideas we could nurture, the possibility of earning an ethical income. I see bees tangoing from flower to flower, fruit trees polka-dotted with juicy jewels, herbs helping heal and the animals soaking up love from cuddling kids.

He turns the letter to me, underneath his smudged, alpaca muesli fingerprint, I see the words: "You have been successful."

I see the vision in technicolour.

I see we – and we all – might have a future.

I see Andrew laughing as I jump up and down like a Rock wallaby on speed.

I see bees, lots of bees.

* * *

NATIVE BEES ARE TRULY AMAZING AND ON PAGE 211 YOU WILL FIND A guide that will help you build your very own backyard bee motel.

BUZZING

*T*welve seconds, that's how long it takes to swallow the big gob of fear chunking in my throat.

Twelve minutes, that's how long it takes to fill in the form to go deeper into debt.

Twelve months, that's how long we have to turn five sloping acres of grass and bracken-covered desert into an inspiring tourist attraction.

I'm beyond grateful we got the go-ahead, but twelve hours after signing for the credit card for our half of the costs, I'm still frozen in front of the computer.

Our office is set up in one half of a single car garage. Thin timber veneer walls it off from the other half which houses the storeroom for guest supplies and cleaning equipment. There's just enough room for us to have two desks and two filing cabinets, and just enough holes in the flyscreen and walls for mozzies and mice to call it their office too. Swat! Andrew greets a mozzie as he walks in.

"Got an action plan yet?" he asks.

"I can see it finished, I just don't know where to start."

"Let's split it down the middle," he says. "You: plant selection, land-scape planning, bee habitat and bees, website, interpretation material, signage, marketing, shop and entry fit-out. Me: irrigation, shop and deck construction, carpark, toilets, fencing, animal petting yards and shelters, gardening muscle."

"But where do we start?"

"Let's get some experts in, mark the place out and go from there."

A month later thanks to garden design gurus Ken, Peter, Elysha and Luke the hillside looks like a giant pincushion. We've dipped hundreds of wooden stakes in paint and positioned them across the five acres. Green stakes mean path edges, white stakes mean trees, orange stripes mean contours, blue means...uh oh, maybe green means garden beds, white means contours and blue means....?

Hole after giant hole is dug in preparation for the more than two hundred plant species which will form the backbone of the useful plant garden. The hillside looks like it's been cluster-bombed and we drag the chicken caravans around positioning them to drop fertility near the craters, above which we plant coffee and turmeric, Chinese jujubes and pecans. One day there will be shade...just not now when we need it.

I don't know if it's the being outside eighteen hours a day, the thrill of creation or the big picture ramifications of the project, but the work begins to give me insane energy. Springs grow on my feet, wings on my shoulders, and muscles appear not just on my biceps, but in my cerebrum too. It's like a calorific overload of the earth entering and channelling the human. My heart beats liquid gold. Everything sparks.

I think psychologists call it 'hyperthymia', when one is so up and energetic it can be annoying to others. Others might call it passionate, manic, loony or 'she's a character', but basically, if euphoria was made illegal right now, you'd have to jail me for life.

I gorge at the buffet Nature sets before us each day, I get fat on it.

Starters – buttery dawns, shimmering pollen, sunsets bronzed like copper.

Main – lemon myrtle's zing, the headiness of Holy Basil, the inimitable musk of buck.

Dessert – the kookaburras' chortle, the bees' chinook, the warming, full-heart glow of a calf at foot.

Supper – milky lips nuzzling fingers, furry chins resting on palms, the sinewy strength of steeds.

Snack – the salty drip of work sweat on the lip, the digestible perfume of loquats, and the schizophrenia of the native midyim berry which doesn't know whether to be aniseed or blueberry.

Every day I feast from a towering croquembouche of energy, set on tables bursting with inspiration, washed down with fountains of refreshing ice. This natural buzz could awaken the comatose, enliven the dead. It feels healthy, but is it?

I don't so much masticate Nature's offerings, as mainline the nutrients through bare feet on clover, apple blossom in the nostrils, honey on the tongue and cicada wings beating circles in my eardrums. With open lungs and delighted heart, I gorge until my senses threaten to tear through and burst the seams of my skin.

We go go go each day, marking out lines for garden beds, digging holes for life, and preparing the ground for the arrival of the old forty-foot shipping container – a metal box a million times more well-travelled than us, having expelled cargo and collected dings in Shenzhen, Singapore, Sharjah, Salalah, Southampton, St Petersburg, Santos, Savannah, Seattle and Sydney. And now, thanks to a towering, karaoke singing, Maori truck driver calling himself 'Mumma', it floats off the back of a truck in little ol' up river Nabiac where it will be converted into a land-locked entry and farmgate shop. From one ride to another.

Splinters, bankers, thirst, hail, 44 degrees Celsius, nails through thumbs....nothing can stop us because we are on the Earth Express, hurtling through space and picking up positive energy and passengers along the way.

We pick up three backpackers; one German, one Austrian, one French: Aphrodite, Athena and Artemis are like three bulging artworks, singlets barely containing their bronzed nubility, colourful tattoos and fleshy joy. As they dig and plant they belt out free spirited tunes, gift rollicking grins and stride proud, sassy and well. It's obvious they are dining – and smoking – from the earth's buffet too.

There's so much work to be done, we pick up more passengers, two strikingly athletic lads from Finland, Aries and Atlas. They can't understand why the three goddesses aren't interested in them, but I know it's because they've already given themselves wholly to *her*.

"It's only 7.30 pm," the boys say as the girls head to sleep for the night. "Why don't you want to party?"

"You'll find out tomorrow," says Aphrodite.

And tomorrow the goddesses flog and flirt and feed those boys into service so that by 7.30pm they too have been exhausted in *her* honour.

We go like this for days, for weeks, for months, for seasons. An energy source burns within, dancing, combusting, fire storming. We are human bonfires pushing wheelbarrows, digging deep, dreaming, growing, testing, exerting, drilling, embracing, sizzling, hammering four thousand five hundred nails into the deck by hand.

The green desert of the grassed paddock transforms under human heat and heart, tiny seedlings begin their stretch toward the sun. Where once was a monoculture of blady grass grows banana, where once was bare dirt dwell Davidson's Plums, where once was vacancy grows valerian, ventricosa, menthe viridis.

She is the oven and we are the cooks. She is the furnace and we are the blacksmiths. She is the crucible and we are the glass blowers.

We sew thin follicles of sugarcane and lemongrass in the hope they'll tuft and thicken to thatch.

We weave in rhizomes of turmeric and ginger in the hope we'll be able to separate and multiply them from fish to loaves.

We knit salvias and abelias, lavenders and pigeon peas, perennial basils and bergamot, in the hope the bees will have warmth in their bellies during all four seasons. Then we embroider arnica, arrowroot, atemoyas; barberry, brahmi, bunya nuts; cork, coffee, curry; daylily, dyers woad, dendrocalmus bamboo; echinacea, elderberry, epilobium; figs and finger limes, gingko and goldenrod, henna and horseradish – and as many of the rest of the useful plant alphabet as we can afford.

We blowtorch day and night to get the job done, to make the dream real, to give sanctity to bee, to plant, to human; to honour our earthen drug lord. The addiction spirals, as addictions do. We go too far. We enrage people when we question their big bills versus their small quotes. We smoke out free riders, fancy pants and philistines. We fry people with our singular focus and inability to chill. We spend more than we should.

We burn and burn until we are extinguished and she is sated. Then she tempts us again: a strip of sugarcane, a cup of tongue-tingling strawberries, a bowl of silverbeet so green it oxygenates. A plate of pink sunset, a serving dish of scent, a smorgasbord of sage.

We sip beers in wheelbarrows as the sun goes down. We shower in dew. We celebrate dawn.

Connections are forged, in the forge.

We perspire. She inspires. We transpire.

Slowly,

ever so slowly,

the earth transforms.

Just as she transforms us.

* * *

GO ON, TRANSFORM YOUR OWN BACKYARD OR BALCONY WITH THE TIPS for bee-friendly plants on page 207.

THE ROVER AND THE CLOVER

I've been running on adrenalin and if there was an award for light sleepers, I'd win hands-down. I hear every scurry, every ululating alpaca call, every imagined burglar. Fighting possums, cable-eating mice, fruit bat flaps. I hear you. I hear the alarm hours before it blares for the 3.30am haul to Newcastle City Farmers Market. Every hour. Awake. Alert.

While Andrew snores exhaustion and exertion, I grab a knife, peer between the blinds, flash the torch around the top paddock letting invisible threats know I'm onto them. Decked out in a ski balaclava and Peter Alexander pug pyjamas I'm ready for anything.

Hyperactive, hypervigilant, hyperventilating. And that was before backpacker Sniper joins us.

Sniper is Irish, whippet-like in movement, darkly handsome. Stubble charcoals his cheekbones and jaw, tattoos fence his arms. It's hard for him to sit still. Our latest Wwoofer is a veteran. Coiled. Twangy like cat gut.

Sniper's survived Iraq, Helmand Province and a roadside bomb but I'm not sure he's going to survive three months at our place. Nor am I sure I want him to. Not because he's not funny, he is. Not because he's not patient with the kids, he is. Not because he knows how to kill people. Um okay, that's why. We're now living with someone who's killed people. On purpose. How come people never tell you that when they apply to come live with you? Then again, I suppose I've never thought to put that on the application.

Name:

Age:

Country of origin:

Hobbies:

Have you killed anyone lately?

When you open yourself to the universe, you never quite know who she's going to send you. And when you're desperate for help, you tend not to hold out for Harvard-degreed humanitarians with references from Amal Clooney and the Dalai Lama.

We're in the first days of Sniper's stay and we're doing the afternoon animal round up. With more than 200 animals, there's always going to be someone missing. We search the flat paddock, along the fence-lines, the overgrown creek, behind the dam. We search the spots where the laggards laze, where the greedy guts, where the Houdini's hide.

Sniper sees her before I do. A teenage Saanen goat. She's down, body still. No rhythmic or reassuring rise of the belly. I'm crunching over dry leaves and twigs, picking up the pace to get to her when Sniper materialises at her side.

He kneels, pries open her white-haired jaws, sticks his fingers in and probes her mouth.

He moves to her ears, caressing them.

Then her muzzle.

I am frozen. I don't know what I'm witnessing but it looks all wrong.

Death is quiet except for the single spectator fly.

He plunges his fingers back into her mouth. "2 hours," he says. "Dead 2 hours."

I cock my head. Unnerved. Eyes quizzing him.

"Afghanistan," he breathes. "Yer find a corpse, feel the soft palette, then the lobe, then the septum. The level of sponge feel gives yer the time of death."

Digesting that.

Thinking about squishiness. Human and goat. The sponginess of death.

"Yer do eet so yer know how long since the enemy cleared out."

The enemy. In the goat's case, is the vampire of the microbe world: Haemonchus contortus. The worm grows up to 2.5cm long, living a cushy existence in the fourth stomach of sheep and goats where it basically does shots of their blood. The female worm is transparent, its festive feed of fresh blood wraps around its white reproductive tract. That's why its common name is Barbers Pole Worm. A downright killer candy cane.

Contortus loves summer rain and the humid coastal paddocks of northern NSW and Queensland. A world champion roundworm, she lays 10 000 mini-me blood-suckers a day. When the goat excretes, mini-me's hang out in the grass until another buffet on four legs cruises along and ingests them again. If the goat's lucky, which ours normally are, you notice she's ill before Contortus contorts her. She'll be slower than the rest of the herd, anaemia draining her gums and eyelids of colour, and the chronic lack of protein leads to a swelling in the throat known as Bottle jaw. Life slows. Death speeds. So do more flies.

"Yer ever put 'em down when they're suffering?" Sniper asks.

"Prefer the vet to do it. Have a gun though if we need to in an emergency."

He eyes me.

I kick myself. I've now acknowledged there's a gun handy. Handy.

"I can teach yer son to shoot properly if yer like," offers Sniper kindly, as though he's offering to teach Jack the piano.

Consternation.

Should I give a veteran a gun? What might he do with a gun? Didn't the famous American sniper Chris Kyle get killed by a friendly?

My brain contortus.

I know that being open to the universe and the liquorice allsorts of humanity has brought our family great insights and joy and helped us stay afloat with the business, but right now I feel it has us caught in the crosshairs. What kind of mother invites strangers, potential killers...real killers... so easily into her children's lives? Then again, is he a killer...or a saviour? After all, when so many wouldn't, he was prepared to do something he believed would keep the rest of us safe.

"Let's get the trailer so we can get her out of here," I say. "Don't want the foxes getting a free feed."

Later that day he repeats his offer to Andrew. Jack at his side thinks it would be cool.

"I'll be with them," says Andrew, "it'll be fine."

It becomes an afternoon activity. Sniper and Jack rifle through cupboards to find a sock big and stretchy enough to fill with sand.

Socks from Target. Target socks.

Then they saw down a culm of bamboo to form a tripod, hanging the sock below to give it weight. A cardboard box with child drawn circles becomes the target. Sniper shows Jack how to etch in the lines that signify his squad. It looks and feels like a craft lesson.

I tell myself, "it's just like darts, but with bullets..."

"Are all the animals out of the way?"

"Yes Mum."

"Are all the people out of the way?"

"Yes Mum."

"Are- "

Aim. Squeeze. Pop. Thud.

"Good one ay! You're a natural mate! But aive some advice fer ya – don't do that movie thing and shut yer eye. Keep 'em both open."

"Why?"

"Yer need to see who's coming fer ya." He tussles Jack's hair. "Don't want to get yer brains blown out ay?" Just two lads chilling. Chilling?

That afternoon as the mountain blocks the burn of the setting sun, we begin digging the grave for the goat.

For every pomegranate, pecan and pear we plant, we drought-proof it by digging a 2m by 2m pit on the high side of the tree. On the low side, we create a mound and plant the tree into it to protect it from waterlogging. The mound also gives the tree's roots the luxury of extra soil before they have to bore their way through the hard clay pan below. The pit itself traps fertility and moisture as it comes down the hill, and is also filled with compost, sawdust and roadkill to give the tree a boost and to act like a nutrient sponge. By dusk, one lucky pecan will luxuriate in the vitamins and minerals of one whole goat.

Sniper digs the blade of his shovel into the soft fill in the pit. Scoop. Shovel. Fling. Scoop. Shovel. Fling. He's on automatic, eyes glazed. I wonder if he's thinking of digging in a desert somewhere. Of digging in. But wherever he is, I'm grateful for his help. The last time I had to bury an animal on my own it was Brutus the old alpaca, it was pouring rain, I was pouring tears, and if you've ever tried to bury an alpaca you'll know it's awkward in the extreme to get that boa-constrictor neck and four lanky limbs to cooperate in the right direction when stiffened by rigor mortis. It was a big and weird-shaped hole, and I ended up with big and weird-shaped blisters. Grubby and grief-stricken I couldn't help but thank the universe at the time we weren't farming giraffes.

Suddenly, Sniper's downed tools. He's crouching, eyes scanning the grass. Inch by inch he moves, palms open to the ground, pawing, sweeping the semi-circle in front of him.

What was I thinking? This must be bringing back memories from the Middle East. Dust. Shovels. Death. Does he think there's an IED? A mine? Is he out of his mind?

"I cain't believe, I cain't believe." He's been holding his breath and he exhales hard.

"Awl my life, awl my life. I cain't believe." He bounces up, grinning like a child. Softening, spinning on the spot, arms wide, fingers tight.

Fingers tight around a four-leaf clover.

"My whole life aive looked for one of these. My whole life."

The sun shines out of his smile. He is a boy. Innocent. Fresh. Uncoiling.

Shamrock magic.

This is why we host, I tell myself. These moments. These moments of pure sparkling joy that make up for the unmet expectations, the stress, the awkwardness.

At dinner, Sniper's smile stays lit and he teases back and forth with the kids, waving his fingers at them and laughing, "get outta my faysssss". It's his signature, a sweet 'back off' saying he knows the kids love and he utters when they get too mischievous and in his faysssss. I settle back in my chair.

I sleep that night. The whole night.

But the next morning he is on his way. Two months early.

"Off to join a feshin' trawler in the Territory."

We thank him, and let him know whatever he needs to do is fine. All the while I'm wondering why the sudden change of plans. Was the work too hard? Did he need money? Are we too boring compared to battle? Did the goat get on his goat?

Andrew drives him to the train for his self-imposed deployment north.

I've never been to war, but with Sniper gone, I think I know what it feels like to dodge a bullet. I also miss him. His electricity. His eccentricity. His ease and dis-ease. At dinner, there's one less pair of legs and elbows to bump at the table. One less laugh. One less person the kids really warmed to.

I prod at my food and an image flashes into my head of Sniper, a four-leaf clover in his pocket, going back into battle. It's just another family dinner, and though I haven't eaten much, I'm not sure how many more of these entrée's into others' lives I can stomach.

I clean up the plates and head into the lounge room. Rosie takes my hand, leads me to the big red chair and sits me down. She lugs over the Atlas of the World, a book almost as big as her, and plonks it in my lap. She wrestles it open to its centre and I see a piece of paper towel. She carefully folds the paper towel back, revealing a perfect four-leaf clover pressed green and flat.

"I've been out hunting too Mum, it's a lucky charm," she says, a sweet, wondrous smile spreading across her face.

I smile back, hug back, feel the urge to give back. I think of Sniper

and whisper under, but with all my breath, what my Great Grand Pa
used to whisper to me:

> *May your blessings outnumber*
> *The shamrocks that grow,*
> *And may trouble avoid you*
> *Wherever you go.*

* * *

TO LEARN MORE ABOUT FOUR-LEAF CLOVERS AND WHAT YOU CAN DO
with them, see page 205.

MOTHERS

We had a goat once who gave birth to twins, she cared for them for exactly eighteen hours, then abandoned them in a paddock so she could go further out for dinner. We had a guest once who was the opposite. A guest who not only stuck so close to her kids we needed only lend the family one set of gumboots for eight feet, but who was so insistent about her maternal instincts, Andrew worried she'd take it upon herself to pour out the formula we'd prepared for the baby goats so she could breastfeed them herself. I hope I fall somewhere in the middle in the mothering department, but tonight I'm not so sure.

The wipe clean chairs of the Manning Valley's Emergency Department are coveted after dark. Pyjama'd kids clutch tear-stained pillows warding off the rants of Jim Beam'd parents; Horlick'd old couples nuzzle together, quiet and solemn; a ginger-bearded man sits straight, kept in place by bandages wrapping both legs to the thigh; there's a stick thin teenage boy with an eyepatch and a girl in his lap losing herself in his other eye; and a cattle farmer – by the look of his boots and skin cancers – hoiking up the lining of his lungs into a moss green tray. I don't join them this time, tonight I'm going through the VIP entry for those who suck at parenting but win at triage. My express ticket: cute kid, gushing blood.

This hospital gets about twenty-five thousand emergency presentations a year and Jack is the latest. The doctor's South East Asian features show no sign of shock at the gaping wound, but his eyes show intense curiosity.

"And...how did this happen?" he asks, peering intently at Jack's gushing thigh...then ever-so-intently at me.

"The thing is Doc", I want to say, *"we've kept him safe from kicking horses, rolling quadbikes, charging bulls, barbed wire and waterways. We've spoken to him from an early age about ice, Facebook and paedos. We've made deals with him that an hour contributing outside buys him an hour of screen-time inside...but-"*

"-He ah, he ah, stabbed himself." So there is no question as to who is to blame, I repeat, "stabbed HIMSELF."

The Doctor, emphasising his raised bushy eyebrow, looks at Jack – raising the suggestion mum is not telling the truth. He wants to hear it from the kid.

Jack nods. "I stabbed myself."

"Are you sure?"

Jack nods. The doctor raises his eye-forest further. "Why did you feel the need to stab yourself?"

Oh great, now he thinks I'm an even worse mother, that I haven't stabbed him myself, but vindictively, purposefully driven him to self-harm.

"It wasn't like that," I say, the words tumbling out fast and defensive, thinking the doctor's thinking Jack's taken cutting to a whole new level. "He was playing Ninja's...for hours...lotsa fun. We were out in the paddock, busy, working, had guests, backpackers, didn't-"

"So, you stabbed yourself? How did you get it in at that angle, in so deep?"

Please Jack, I'm wishing, *just tell it how it was, don't get shy and bottle up, don't clam...and please don't go into shock – I don't want to be walking outta here in cuffs!*

"I was Ninja trampolining, somersaulting with it, and when I landed, it just stabbed in." Yep, a six-inch Shimano fishing knife. Deep in the thigh, a one and a half twist at velocity. "I didn't do it with my machete though," he said, as though that sounded better.

Apparently, from the eyebrows bushed in my direction, good mothers don't let their kids play with knives or machetes. And they especially don't let their kids play with them on trampolines. Apparently, bad mothers do. Apparently, my free range motto: "Wrap the kids in cotton wool after they have an accident, not so they never do", makes me a bad mother tonight.

The facts – having assured the doctor of my inattentive parenting skills rather than malice – lead him to focus on the bleeder.

"You a tough Ninja?" he asks. Jack nods. "Good, because we are going to need to stitch that right up. You're very lucky you know, you stabbed yourself in the safest place." Then he looks at me. "If you were in Papua New Guinea, this is precisely where they'd spear you to teach you a lesson but not kill you."

On the drive home, Jack sleeps thanks to the hospital-issue painkillers, and I'm wide awake churning over the Doctor's coded lesson for me, basically: your kids are more important than your work, your guests, your helpers, your farm. I glance at Jack through the rear view mirror, realising that though we started this place to make a difference in our kids' lives and for their kids' kids, it doesn't make it the right thing to do it so intensely.

I vow to try harder, to spend more time with him, to encourage his friends over, to activate his space, not just be around.

A few weeks later we get the chance. A new boy started in his sport team a month ago, and we've invited him over for the afternoon. Being out on a property, it's hard for Jack to fit seamlessly in with the townies. And not being born in the area, he doesn't have the social benefit of the kids whose parents knitted so tightly in mothers' groups they wear matching jumpers. It's basically up to me to make sure this social foray goes well.

Jack runs off to play with his new mate Dylan as soon as they arrive, and I'm thankful I have a whole farm to small talk on with the parents, not just a suburban entry foyer.

"Hi, I'm Jack's Mum," I say.

"Hi, I'm Reese, this is my husband Leo." *Wow*, I think, *what cool names! Beats boring ol' Anna and Andrew any day!* I'm in my pig-tail that hasn't been cut in a year, grimy jeans and can feel dirt grains on my face, so it's hard not to notice her decade-younger smoothness, perfectly applied eyeliner and recently balayage-d hair. Hubby Leo is gym-buffed and looks dyed too. *Hair really matters to them*, I deduce, but hey – it's our kids who want to be friends, I'm just happy for them not to think we're serial killers so the kids can hang out. I lead them down to the animal yards, and let the small talk begin so they can entrust us with their son for the afternoon.

"Where have you guys come from?" I ask.

"Big farm, up Queensland way. Thousands of acres," says Leo.

"Wow! This place is only ninety, we'd be your veggie patch! What brought you here?"

"Is that a harness on that goat?" asks Reese.

"Kind of like a maternity bra," I explain. "So, did you have sheep up there?"

"We rented a house on it, didn't work it," answers Leo.

"Cool! What did you do out there?"

"And what breed is that?" interrupts Reese.

"Anglo-Nubian with a bit of Toggenburg thrown in. Creamy milk. So, have you found work down here yet?"

"Do you sell the milk?" she asks.

"No, it's illegal to sell it raw, but we drink it, make paneer, use it for soap. So yeh, what do you do for work?"

"Sales," says Leo.

"And mothering," says Reese. "We've moved around a bit for his work, this is Dylan's fifth school."

"That'd be hard getting him settled in," I commiserate.

"Is that an alpaca or a llama?" asks Reese. And I realise she's as uncomfortable with the small talk as I am, probably on the introvert side too...or, is she...is she diverting?

Something's niggling at me, maybe it's all the adventure books I've read, maybe it's all the movies I've seen, maybe it's their cool names, their dyed hair. Maybe it's because I have a wild imagination, or have become a heat-seeking missile for authenticity. Maybe it's because I'm a terrible mother who lets her son play with knives and machetes and just can't shut up and secure him a friend. Maybe I'm just nervous and want to make a joke, maybe I just don't think at all. Maybe I just instinctually take the small talk to big talk.

"You guys on witness protection or something?"

The wind stops.

Bravado the alpaca cocks his head.

The bees fly slow-mo.

The world turns, ever so s-l-o-w-l-y.

Shit!

Leo and Reese suddenly look like Buster and Busty...and busted.

Backpedal, backpedal, backpedal.

"Hahahahahaha. That's the running joke at the post office, that Andrew and I are on witness protection! Hahahaha!" I slap my thigh for good measure, "It's because of all the backpackers, we get letters addressed to Giovanni's and Luisa's, Katerina's and Gilles...either that or they think we're international fraudsters. Hahahaha. That's why I mentioned the witness protection, Hahahahaha. So...."

So, the afternoon get-together goes downhill faster than a cow with bloat. Her heavy makeup is no match for the colour draining from her face and I can't help but notice they engage in a blood transfusion, the red rising in his cheeks till his whole face boils like passata on bottling day.

"I'm sorry," I say, and I mean it. I'm standing in an animal petting farm, stuffing up another person's mothering while trying to improve my own. I have never been sorrier in my life, for these people, for Jack. In the movies, the witness protection people always get tracked down by the bad guys...so what does that make me? I feel like the farm's become the Bermuda Triangle of the North Coast, a vortex for mothering missteps and disappearing parenting skills.

Within a week, they've changed their mobile numbers, within two weeks their hair, and two months later Jack's new friend has moved on again. We don't tell him why of course, we just commiserate with him that he's lost a mate.

"They didn't have to admit it," says Andrew to make me feel better. "They could have had a better backstory, plugged the holes a little.

And maybe I could have plugged my mouth.

I seek outside guidance on modern mothering and start watching Housewives of Beverly Hills/New York/Orange County – it doesn't matter the city, they're both the shock and absorber I need to see how other mothers do and don't do it. When the kids and backpackers catch me watching the shows I tell them I just want to know what the enemy are thinking. I learn a lot about wealth, I learn a lot about grown-up tanties and I learn that my four pairs of shoes – who I am incredibly loyal to seeing they get me through every situation – gumboots, hiking boots, Crocs and a mid-heel, don't even figure in the other housewives two-hundred room, shoe-hotel cupboards...but I don't learn much about parenting.

So I start absorbing what our farmstay guests do. There's Christine who loves with big heart, true smile and a daily supply of fresh-baked treats. There's Lisa who loves like a golden retriever: gently guiding, alert and so kind. There are brittle mothers trying to soften in yoga pants, mothers treating their kids as limb and social extensions – not separate beings, mothers 'doing the farm holiday for the kids' and begrudging every moment spent out in the country air even though they admit their kids are the happiest they've ever seen them. There are pamperers, self-esteemers, beraters and mothers who just send the nanny in their place. There are mothers I wish were my own and mothers our kids wish were theirs. So many types of mothers, so many ways. None of us know what we're doing but we're all trying our best anyway, whatever the shoes we wear.

I'm still daydreaming about being a better mother when I hear shrill screams from the pool. They're screams of fear, blood curdling and panicky...the sound I made once when I saw a brown snake swim across it.

"What's wrong, what's wrong?" I run outside and to the gate. The mother guest, a business psychology executive, stands at the side of the pool with her two daughters. I search the sparkling water desperately for a snake, a drowned ringtail possum, a misfortunate chicken, the first crocodile to ever get so far south, but it's crystal clear. The kids continue to gesticulate madly at the water. The mum takes their hands and steps them back from the edge. The screams turn to whimpers.

"Oh, don't worry," she calls to me brightly, "we had a nanny I didn't trust once, a house with a pool – it was all so risky I gave the girls a clinical fear of water. Now they're old enough, thought I might reverse it while we're here...better than the neighbours at home hearing the screaming. You know how it is!"

It always takes a while for my brain to process things I'm not expecting to hear. And I've never heard of that. I nod, smile and retreat to process the planned, scientific psychological manipulation of children as a parental tool...and why if she didn't trust the nanny, she didn't just choose to find a new one? Or if she didn't trust the pool, why didn't she just install a better gate? The encounter swims in my head, waves of disbelief crashing about, judgement wanting to come up for air. But I snorkel it down, horrified, but delighting in the idea of inducing a few clinical fears into Jack's head – starting with blades, mates with gangster parents, and moving on to teenage girls and fast cars.

This mothering thing is threatening to blow my mind, so I put a lid on it with another dose of Housewives. I'm feeling better about myself already. The problem and benefit, I realise, of experiencing so many different mothers at the farm, is it's making me experience me.

* * *

WANT AN AWARD FOR MOTHER OF THE YEAR? TRY COOKING THIS crowd-pleasing paneer! It's on page 229.

SEXY BEASTS

*S*ome people take on the challenge of half marathons and Tough Mudders. Some take on Everest and the Seven Peaks. Me? Right now, there are two challenges I want to rise to: One: I want to actually *want* to have sex again. Two: I want to find a place to have it where I won't get caught starkers by backpackers, guests, neighbours, kids, or animals with stalky eyes and snakes with staccato tongues.

It's a challenge that feels insurmountable, as unmountable as I seem to have become. It's nothing to do with Andrew, because if I was even wanting an affair, I'd still be having it with him. It's just lately I'm not wanting – at all – and there are three reasons why.

First, I have to get past the speedbump of breasts that feel more functional than fond-lable, that's what breastfeeding did to me. What were once voluptuous, tingly, oh so sexy to me and man swellings, have morphed in my mind to little more than factories to produce tepid, on tap milk. And though I no longer breastfeed, being surrounded by hairy goat udders every day who do, rams home the idea of breast as tool rather than turn-on. It seems when I freed the nipple to milk it, my libido and lust made a run for it too.

Second, my mind is where it's all happening, there's an orgy going on up there. An orgy of who's, what's and where's; insight, oversights and strange sights; dreads, dreams and day to day. Much ado up there, means nothing much left for down there. There's room in the king bed for discourse, but just a poky mattress on some far away floor for intercourse.

Third and most overwhelming, there's the seven-year battle with my delivery chutes. Bowels that spit tiny rock cakes – next to a flat-bread-rising column of yeast – guarantee me endless appointments with gynaecologists and colonoscopists and dear white porcelain friends. Finally, someone suggests it might not be childbirth or polyps or weak muscles or anxiety or tight-fitting jeans causing the problem...it might just be my genes: HLA DQ2 and HLA DQ8. Yep – Coeliac Disease.

"Great, so you just need to stop milking the goats, stop thinking, stop eating wheat, and we can make cake!" says Andrew, ready to feast after dining on patience and kindness for way too long.

And maybe I just need some sleep, a full renewing week of slumber undisturbed by market alarms, hooting owls, feral cats, pacing kids, bleating goats and guests or Wwoofer raps at the door.

It's as though now I've pinpointed all the problems, there might be a solution. Within two weeks of gluten free my columns normalise and I start sleeping better too. Within another week my brain blanks enough to reach for a lacy bra. Bugger, then I get my period, will have to wait, sorry Andrew.

But I spend the week well, engrossed in nature porn.

I watch male heads fall from the sky, severed from their penises during their first and only copulation. It seems Queen Bees are intent on diversified sperm, not company, and when they successfully mate in the mile-high club, the male drone bees lose not just their virginity and frequent flyer miles, but their penis and abdominal organs too. I'm not sure that's the thrust of the birds and bees chat I heard as a kid.

I witness Harley the billygoat curl his lip and use his extendable firehose to urinate on his own beard, a stinky yellow tinge of after-shave that has the ladies escaping chain-linked, electrified and barbed wire fences to be with him. He's like the open-shirted merchant banker at the bar, doused in Tom Ford Noir Extreme, 'capturing the exact moment when a man reaches extreme new heights: his sophisti-cation becomes magnetic, transfixing the world at large' ...or at least this bevy of does who are amid twelve to thirty-six hours of raging oestrous find him so. Later, when raising his quadruplets, they'll probably wonder what the hell they saw in his shallow, smelly self and dyed yellow locks.

Then, I'm leading a tour group of seventy-year-olds on a farm tour when my speech about the importance of pollination is interrupted by group sex. The tour goes from PG to R faster than a fourteen-year-old on the Internet. The ensuing humping, humming, alpaca gangbang entertains for the next forty-five minutes of their visit, even with buckets of food, I can't tear them away – the tourists or the alpacas. I explain to the group that an experienced female alpaca knows she is pregnant within four days of coitus, and the way to confirm is to bring the male alpaca back in with her for a 'spit off'. If she sits down she's happy to go again, but if she spits green, half-digested gobs at him there'll be a sex drought for the next eleven and a half months of gestation, and a few more months while she lactates.

"We've got it pretty good then," says a bald guy to the other men, his wife too slow on the walking stick to whack him quiet.

And then ManCat arrives. ManCat is the latest American back-packer to join us. He's got a degree in ag, his parents run a nursery in the USDA's climate zone 10b and he seemed like such a catch when he emailed us. We set him up in his own cottage rather than bunking him in with the three backpacking female Brits who are already ensconced in the other cabin. We bring him fresh towels, sheets and plump pillows and settle him in, he's just come from an Ashram so is used to much simpler accommodation. We figure he'll be happy there.

The next day Andrew and I are so ready! He's milked the goats, the kids are off to school, the guests don't check in till two pm and the

Brits are beavering away on the hill. We wash our hands at the tap outside and run to the house like teenagers; under the awning, through the kitchen door, into the lounge room, and right into ManCat who is curled asleep – a huge hairy adult foetus – between us and the corridor to our bedroom.

Passion pops.

ManCat apparently likes to work early and nap through the day, not in his cottage or under a tree or on a picnic rug down by the dam, but around the leg of our lounge room table, or stretched tummy-fur up in front of our bookcase. Basically, in the middle of our path to lust. Awkward.

It turns into a very long day, and with the guests arriving late, Andrew and I make plans for a morning rendezvous.

6am: We begin to warm, to writhe, until RAP RAP RAP – the new guest children hammer at our bedroom window.

"Are they in there?"

"I heard something. They must be."

"Come out! Come out! Farmer Andrew! Farmer Andrew! Mummy says you need to stop the rooster crowing."

"Are you in there?"

"He's in there, I heard him."

"Farmer Andrew, Farmer Andrew!"

Hormones get hauled out of bed. We don't get a chance to f***, but I say it!

And of course, our own kids that night…they fever, they need help with homework, they are having trouble with friends, they don't want to go to bed, they aren't tired, they want to fall asleep with us, they have a dream they want to share, they forgot to brush their teeth, they have a cramp.

The drought continues.

Life piles up. Guests pile up. Things pile up on the kitchen counter. ManCat is not just a napper, but a fermenter. Bowls and jugs and cups juggle for space on the bench housing contents days into decomposition. Raw goat milk curdles by the jug-full waiting to be lapped, green slime runs it tongue over vegetables, there's a soup of microbes mewling next to the sink. Do I throw it out or is it his next week's gourmet lunch? Is that carcinogenic or a perfect kombucha? Are they pickles or is he pickled?

When ManCat does fermenting, no matter how hot Hollywood stars make it sound, and no matter how good it is for you, when it's all hissing, moulting and scumming on your kitchen surfaces, it is just not sexy.

But Andrew is and I want some of that.

The afternoon is torture while I wait for him to finish the activities with the guests. Then it's the big dinner table chat with the Wwoofers, the job list, the wash up. Then it's time to get our kids to sleep. They are not budging. Lucy wants to keep reading in her room. Rosie is playing in the hall and Jack is on the chair next to us saying, "Please, please, please can I just stay up a little longer?"

"Come on, it's bed time. Come on guys, it's been a long day. Come on you three!"

The tension is building but I know a big argument now will escalate, then destroy the vibe completely. I don't know where it comes from, and I can't stop it as it escapes my lips, but out of my mouth comes these words:

"Jack, darling, please go to bed – Daddy and I want to have sex on the couch."

Jack jumps up like he's been stung by a wasp, the massive-winged, deeply disturbing, incredibly penetrating taboo wasp. He looks at us in horror, he fumbles for words, but all that comes out is, "nooo, noooo, nooooo."

Then a shocked smile starts to creep across his face and he races down the corridor to his siblings "Nooooo Noooo!". We hear giggles. The corridor door opening and closing, opening and closing. More naughty giggles. "They wouldn't!" "That's so disgusting." "That's what she said." "Eeeeewwwwwwww." "Eeeeeeewwwww." "Noooo. Nooooooo." "Tell them that's so no! Just so no!"

We hear doors opening and closing for the next ten minutes, more giggles, more outrage, more "eeeeewwwww's" more "noooo's".

Andrew and I laugh so hard our eyes water and overflow. I can't remember the last time we laughed like this, felt levity like this. Who would have thought 'parental couch sex' is the only phrase parents across the world need to get kids out of your face and off to bed! Unfortunately, we laugh so much and for so long, we kill the sex vibe all by ourselves.

The next day we really gotta get ourselves some!

I carefully zip my naked self into my beekeeping suit and prowl up the driveway looking for Andrew who should have just finished morning activities with the guests.

"Can I come beekeeping with you?" asks ManCat, who's also out on the prowl.

"Oh, um, not today, they're a bit, ah, fired up so I better go on my own. Tomorrow? Okay? Tomorrow?" He's a great guy except for the awkward napping positions, and preference for off foods and long stares, so I feel a bit bad for lying, but I'm on a mission and there's definitely no room in my lap – or corridor – for ManCat.

I hear the clinking of barrels in the feed shed, it's Andrew making up extra muesli for the newly pregnant alpacas.

No one is about so I half unzip, giving him just a flash of the taut flesh within the baggy whites. His eyes light up and he peers outside the stables, scoping the farm.

"Quadbike? Creek?" He asks. I nod, down by the creek there's the perfect private spot, dappled in sunlight and hidden from view.

We jump on.

"Farmer Andrew! Farmer Andrew! Can we ride too? Can we? Can we?" Chant the guests who have come running at the sound.

"Not this time," he says, revving the engine, "Anna and I have something only we can do."

We gun it out of there and we do. And we do. And we will. It's taken a while, a long while…an extended ferment really. I didn't really think it possible, didn't think it would materialise, but my health and thirst are back and now I'm the one needing the quenching.

There is just one lasting casualty it seems; the kids don't seem to want to sit on the couch. But hey, that's no great loss, it tends to make even more room.

For the two of us.

* * *

IF YOU STILL HAVE AN APPETITE FOR SAUERKRAUT, IT'S VERY GOOD FOR you by the way, there's a recipe on page 231.

CHICKEN

oday I need to kill a chicken. It's just a chicken, but I'm chicken.

If Andrew were here, he'd take care of it, enabling me to be at arms' length, but lately he's been somewhere else whenever someone is about to die, needs help dying or dies inappropriately. Like today.

Her days of modelling on the cover of egg cartons might be over, but even in old age, she's bright-eyed and her chuckling cluck still buoyant with banter. But sometime in the last week her coppery feathered right leg stopped working, and now her body twists and strains with the effort to drag it along. She can only move an egg-length at a time before she tumbles down. I don't even need to call in the kids to catch her, I just scoop her up.

Up until a few days ago she was still living the high life at our Hens on Holidays Caravan Park, now a cluster of old caravans rebirthed as mobile hen houses. Painted bright orange, lime green and electric blue, the caravans provide fox-free night time accommodation and a funky place to call home after a day free-ranging.

Her Boho gypsy days have suited her. I love how she feels tucked up in my arms, she's like a hot water bottle full of trust, a calm ripple against my body. I gently stretch the bad leg and finger the spindly bones from her toes up to her thigh, checking for bumps or ridges or cracks, she doesn't even flinch.

"That's not good," I say to her. "If there was a break we could splint it with a paddlepop stick like we did for your old friend peacock."

I carry her with me, not wanting to put her down to be attacked by ants she can't turn fast enough to peck.

We head into the house. It's the first time she's been back inside since emerging six years ago from the incubator nursery set up in our bedroom cupboard. Back then she was a fluffy marshmallow of sunshine, tottering about like the star of a tissue commercial and mesmerising us with her visual and vocal avian adorability. Cheep-cheep, cheep-cheep.

I get an old blue towel, fold it four times and lay it in a cardboard box that once held oranges. I pop her in and we go sit down in Dr Google's waiting room.

I type in, 'chicken leg paralysis'.

Broken bones, no. Temporary paralysis from predator attack, no. Malnutrition, not with all the treats she gets. Botulism, unlikely. Tumour, possible, humane euthanasia recommended. Marek's disease, possible, isolate from flock and humanely euthanise.

I swallow. It's one of those back of the throat, uncomfortable situation swallows. The swallow you do when your brain realises things are about to get hairy. The swallow you do when you realise you need to face something you don't want to, the swallow you do when you don't know if you can.

Euthanasia: the painless killing of a patient suffering an incurable, painful disease. Mercy killing.

Chicken a'la Orange and I head back into the sunshine to discuss her end of life plan and I notice her breathing's more laboured than yesterday. Her tiny little mouth and nose focused on sucking in life to deal with pain.

"It's not looking good," I say as we settle in the shade of the lemonade tree. It's one of her favourite haunts, a great spot to catch bugs who are lured to the juicy yellow fruit and fleshy green leaves. "You can't go on like this, you're getting all mucky, the other birds want to peck you. You can't live in a cardboard box the rest of your life. And you might have a disease that can kill all the others. I don't want you to be in pain." Chook blinks trustingly. "I'm sorry, I really am. Look, I'll just go and get you some food and water."

I fill up her water and seed, then sit down in Dr Death's waiting room. I type in, 'how to kill a chicken?' I mean, I've never done it, that's Andrew's forte, but I do know how to kill a chicken, you just cut its throat or break its neck, or axe it or shoot it. But I don't think I can bring myself to do any of those. They're all way too personal, too hand to hand combat, too many chances for me to stuff up, to just turn her into a chicken quadriplegic in a wheelchair. And I don't want her last moments to be Texas Chainsaw Massacre, long gory schlock and the awful twelve to twenty-five seconds of consciousness after the cut. I don't want her to suffer...which is why I need to kill her, well. I need a creative solution.

I want to stun her first, then she won't know what's coming. I need an electric current...our electric blanket's faulty but I'd probably kill us both if I wrapped her in it and turned it on. A cattle prod might work, but we don't have one of those. Hmmm.

Chloroform! A tiny little hanky over her tiny little nose? But where can I get some of that?

Phenergan! That stuff hidden in the top cupboard we once used guiltily to squeeze a medicated night's sleep out of the kids when they were still roaming the house at midnight. I check. Damn! It's out of date...but maybe I could dropper it into her mouth or encourage her to do shots, lip-sip-suck, Phenergan beer bongs even! And once asleep I could lie her across the drive and run over her with one tonne of ute. Fast and hard. Then reverse.

Erk. Too brutal, too messy. No chicken will ever want to cross the road after that.

Or maybe I could make up a do-it-yourself carbon dioxide chamber! All I need is vinegar and baking soda and I have plenty of those in the kitchen cupboard...mix them up, get a tube, a cover and gently gas her in the big punch bowl?

But is that the way for a chicken to go to her final sleep? In a punch bowl? And could I ever serve sparkling passionfruit tropical punch out of that bowl again? It'd be too Ghost of Christmas Chook.

I realise my breathing is fast and shallow, oxygen barely skimming my lungs then rushing back out. Skim, surge. Skim, surge. For the life of me, thinking about ways to kill her, I just can't seem to draw in the breath I need.

Why am I struggling with this? She's just a chicken! Australia's favourite dinner. She's just a chicken, a domesticated, small brained bird whose biggest thrill is scratching in dirt and finding worms. It's just a chicken. But I've read the science papers that explain how chickens have a sophisticated language, that they have at least thirty different calls that can communicate everything from "hey get over here this food is awesome", to "get the heck outta here, there's an eagle overhead." I know they talk, like me, just a different language. Over the years, I've spent hours watching her and the flock interact, different personalities, different races... Light Sussex, Araucana, Isa Brown, Australorp, Barnevelder, Silky...whites, blacks, browns, greys, mottled, speckled, frizzled. I didn't need another scientific paper to tell me they feel stress when they see one of their own injured or stressed.

They feel. But she's just a chicken, and I'm just trying to do her a favour so she won't have to grovel in the dirt in pain, and be stuck stationary to be eaten alive by ants and maggots. She's just a chicken. And so am I. I don't have the courage to do it. To help. I'm not as humane as I think I am. As I'd like to be.

I want a clean, tidy and painless death for her, a hospice unit where she can be morphined up to those tiny little eyeballs, the calming beeps of the heart monitor, the liquid drip in the arm suggesting all will be okay if she's just hydrated...until the last gentle minute when she slips away. Peaceful. Taken care of.

But she's not slipping away. Her eyes are right now boring into mine, so alive, so alert, even as her body sags. I have to do something, I can't let this drag on. I head back into the office, activate the dark web, and type in, "chicken assassin". But no luck...the only results it turns up is for an internet game titled "Chicken Assassin - Master of Humiliation".

Humiliated, that's what I am. Ashamed that I can't give this bird a painless, dignified exit. Or can I? Damn it! I really do need a professional chicken assassin. That's it!!!! I dial the number.

"Hello? Um…can you tell me how much it costs to put a chicken down?"

"I'm sorry, did you say, a chick-ken?"

"Yes, a hen actually."

"Well I'll have to find out, I don't think we've ever been asked that before, dogs and cats, all the time…but chickens…" She covers the phone with her hand but I can hear her whispering to someone in the office. "A chick-ken!"

The phone feels sweaty against my cheek, money is tight but right now I don't care what it costs, I haven't spent money on myself in ages and putting this chicken down nicely suddenly means more to me than any piece of clothing, any book, any dinner out.

"Vet Pete is doing rounds down your way this afternoon, if you'd like him to come see your chicken we can do it for the call-out fee."

They must be laughing in their office now, a farmer having a vet come out to put a chicken down. A fool farmer, more money than sense.

"Book us in," I say and go back outside to sit with Chicken a'la Orange.

She's just a chicken, just one of eighty on the farm. Why do I care so much? Have I been infected with anthropomorphism from all the Disney movies I've watched? Don't I realise that Bambi is cute for a reason, that Simba was sculpted to have us roaring for him, that in any other setting Jiminy Cricket would just get sprayed or squished under a boot?

I stroke her silky feathers and Chicken a'la Orange clucks gently, she's no cartoon character, she's real. The sun warms us both.

I don't think she feels exactly like me. But I do think she feels. I position her in the dappled shade so she won't overheat and head into the office to tap out emails rather than think any more about tapping her on the head.

I make the mistake of clicking on the news rather than the inbox. Here I am going out of my mind thinking about how to put a chicken down humanely and ISIL is burning humans in cages. Argh! Now I feel guilty, the money I'm spending on the vet, such a privileged choice, could probably save a human life overseas, make a real difference. But where does mercy start? How does it end? And to whose end? And someone somewhere will be telling me what a waste to put down a perfectly good chicken without eating it. What is right? What is wrong? I click off the news and kick my chair away from the desk. I need sunshine.

Back in the garden, I tempt Chicken a'la Orange with some fresh parsley and she scoffs it like there's no tomorrow. I cringe because I don't want to tell her she's right.

Thank goodness for the distraction of the vet's car pulling into the driveway, the bent gate being hoisted open. Vet Pete's pretty much a regular now having been here to hoist out twin lambs, grease the intestines of colicky old thoroughbred Hollywood, lop anaesthetised ball-sacks from wannabee bulls and tend to the old and infirm. He has a gentle, professional manner that puts you at ease with disease.

"Hi Anna, this is our fourth-year intern," he says, introducing me to his sidekick in the matching King Gee overalls, overalls the darkest shade of blue so as not to show off the accumulated daily goo. "Apparently, it's something about a chicken?" He doesn't emphasise the chick-en like the receptionist, he's the complete professional. "Is this her?"

Yes, this is the only chicken in a cardboard box in the garden.

He picks her out tenderly and I think of all the breast tenders in the supermarket. His two hands lock around her like a safety harness, he rocks her gently from side to side. He inspects the colour of her comb, nice and red. He looks at her eyes, they shine right back at his. He begins feeling the top part of her thigh, the treasured drumstick in most boxes of KFC. Then finger feels his way down her leg to her toes like I'd done.

"No fractures that I can feel," Vet Pete says and hands Chicken a'la Orange over to Kid Vet.

"Me neither," Kid Vet concurs after a quick feel.

"So…" Pete says, "what are some other possibilities?"

Kid Vet stretches out the wing on her good side and Chicken a'la Orange protests by flapping it.

"A tumour?"

"Possible."

"Marek's disease?"

"Possible. But we'll need samples and an autopsy to confirm. If it's Marek's Anna, you could lose your whole flock."

In my mind, I see the A-list coroners and medical examiners of Hollywood coming together for autopsy a'la Orange…the crew of Castle, CSI, Law & Order crowded around a small silver serving tray, even old Quincy and Dr Kay Scarpetta there trying to work out why the sky eventually fell in on her.

"How would you put down a chicken?" Vet Pete asks JuniorVet.

"You've never put down a chicken?" I ask, but he turns away without answering, his ears listening to the student.

"We need a muscle," says Kid Vet. "A needle into the liver is too slow, painful. With the last vet I interned with, we found a vein."

"Really?" says Pete quite interested. "Do you want to be here for it?" he asks me gently.

I nod. I mean, it's not like I can call in the death sentence and not be there when it's carried out, that would make me like the farm version of The Pentagon and Kremlin.

Vet Pete clicks the latches and opens his black briefcase of tricks.

I can't hold her hand so I hold her eyes. Right until this moment they have been wide open, taking everything in, meeting the world. But at the sound of the briefcase opening, and my quiet agreement to the take down, she closes them.

She doesn't open them again. Not as we tap her leg like addicts trying to get a vein up for a fix. Not as Pete draws the green dream into the syringe. Not as I gently stroke her head, or when I get a sudden moment of clarity and urgently call out, "can we do it in the paralysed leg so it doesn't hurt?"

"Good idea," says Vet Pete.

She keeps her eyes closed. Firmly. And I notice. I see. And I start to tingle. It's a physical reaction to my realisation that Chicken a'la Orange's brain and understanding of the universe is bigger than my own. The closing of her eyes at the closing of my door on her, doesn't feel like a coincidence and it creates invisible currents in my body. We do not know, what we do not know. But she does. She doesn't open her eyes again, and not because she is dead.

She is blocking us out, taking back the little left time for herself. I wonder what she is seeing. Her family and friends? Green pastures and sky so blue you could swim in it? The time she just avoided the fox? Being held gently and caressed by enamoured human children? Rocky the rooster bringing food to her and dancing enthusiastically till she ate? The freedom of pooping anywhere she liked and just walking away? Was she recalling Andrew's smiling face when he opened the door to her caravan each morning? Or feeling the rush of air under her wings as she flapped out through air so fresh you could sell it? Is she savouring memories of hot summer days under the cooling pecan, or the warmth of her cousins tucked on the roost next to her on cold winter nights?

It was so obvious she was seeing something, I just wish I knew what it was.

I don't know what I believe about life after death, or nothingness, or reincarnation, or energies, but right here right now there is a gap in the garden, a gap in reality, a gap between the logical and spiritual. I step into it. Feel the energy. I feel my eyes go big, then orb, then soften to jelly.

The magic milk takes over, now it's her whole body that is limp and I realise I'm holding my breath.

"Did you notice that?" I say to Vet Pete, "how she closed her eyes when we were talking, when you got the syringe? Did you see?" He nods. Nods at the crazy lady? Nods because he agrees?

"It's more likely it's a tumour than Marek's, don't worry about it," he gently steers the subject to safer ground.

I shut up except to say goodbye.

The copper of her feathers still shines, but she no longer does.

She's just a chicken. She's just something other. But she's single-leggedly breathed fresh life into my feeling that every life is sacred... every colour, every religion, every creature, every river, every ocean. And definitely, forever from this day, every chook.

* * *

FOR A CHICKEN BURGER RECIPE MINUS THE CHICKEN, SEE PAGE 233.

DECLUTTERING

*I*t's time to declutter. Less than a decade ago we could fit everything we owned into a two bedroom, inner city apartment, but now we can't stuff our accumulated belongings, offspring, ideas, backpacker entourage, animals, produce and business into a whopping ninety acres! So much for the simple life.

I'm at the huge timber lunk of a thing that is our dining room table, it can easily fit eight around it but I'm sitting here alone and cramped, hemmed in by the current production line that is the farm. Tiny paprika and rosella seeds I'm saving to re-plant next season dot strips of paper towel. Spread out on six, long-as-your-arm, commercial baker's crates, is the fresh cut herbal foliage of in-your-face-fragrant rose petals, rose geranium and lemon balm, ready to be dried for use in tea. Eight crates are stacked nearby, they stand as tall as me and house two hundred and forty bars of goats' milk soap. They're four weeks into the curing process and the scent of honey and organic oils of lavender, lime and mandarin crowd for my attention – and will continue to for another two weeks until they're at their firmest and mildest to sell.

At the far right of the room are cartons of last month's delivery of 1 000 tea light tins and their wicks, flaring at me to be filled with the beeswax that waits stacked in tubs under the table. Perched atop the carton pillar are twelve beautifully-patterned, op-shop salvaged china tea cups awaiting their own candle rebirth. The problem with producing, I'm finding, is there's always production, always something needing to be done or needing a place – at my place.

It's drizzling outside so sheets from the guest cabins and table cloths from the bus group morning tea are triple-draped over doorways to dry. As usual, there's more linen than doors. I hear Andrew rattling pots and pans in the kitchen. Plates being stacked, more plates, more plates. Fifty tea cups from the bus group morning tea. A cascade of cutlery.

I look to the tiled area leading to the kitchen where size six lingerie, teeny floral shorts, skinny jeans in every shade, tank tops, jackets and active wear spill out of the two laundry baskets on the floor. The clothes aren't even mine or the kids', they belong to the female half of the latest Wwoofers to join us. The Maple Leafs only arrived this morning and like the kombuchas brewing at the other end of the table, I have a feeling they're going to be an acquired taste.

Then a *thunk*. The Shetland ponies have escaped their paddock again and are staring at me through the lounge room window. I tense, ready for the film crew from Hoarders to burst in and capture the whole scene, but it's only Andrew.

"Dammit! This place is out of control," I spit as he picks his way past the lounge room obstacles, his jacket follicled with damp white goat hair, his face a picture of *I am so not getting into this right now, I have my own shit to deal with.*

My gaze turns to the unruly piles of reading material teetering on the corner of the table, threatening to spill pages of secrets on mead-making, humanure and farmhouse cheeses. I look at the book skewed atop the pile of other unread tomes, Peter Walsh's book 'It's All Too Much'. I randomly flick to a page and read, "It's about keeping things that make sense for your life – your real life, not a fantasy of what was or what could be."

Thunk. Thunk.

"Get away before you smash the window," I yell at them, then whisper so the kids can't hear, "bloody Shetlands."

I've never thought animals could feel like clutter, until staring back at these two equines right now. The responsibility of looking after them, cleaning up after them, chasing them down, finding the money to care for them is suffocating.

When we bought the farm, the mum and son Shetlands came with it. Bronze-age equines of stocky, sly heritage, we found out just after settlement they also required bi-monthly visits from dentists, vets and podiatrists. That would have been palatable, just, if they were like Fizzy or Ribbon from My Little Pony, but Cindy and Cody rule the paddocks like malevolent garden gnomes, nipping and chasing anyone smaller than themselves... sheep, goats, toddlers. When it comes time to corral them for guests for hand-led (or is that towed?) pony rides, it takes half a morning and minimum twenty dollars of hay before they deign to be saddled. Then they extort another half bale to take a measly step. They turn their butts to me, they glower, they send me bills, but they've also sent me the joy of watching Lucy get the better of them, her youthful determination to not be thrown, her will to work with their ways. Recently she's grown taller than them, but Jack and Rosie don't have the temperament or death wish to take them on. I don't blame them, I'm beginning to realise sometimes taking things on just isn't worth the ride.

I don't have time to chase the Shetlands down right now, I've got to take on dinner. I navigate out of the lounge room and into the kitchen where four more grey baker's crates are in use for plate and cup drying, it's a high-rise cityscape of crockery and it takes up the bench space I need to prepare dinner. I grab a cutting board, bridge it over the sink, the only spare space, and begin chopping.

Carrots. Broccoli. Capsicum. Snow peas. Baby corn. Bok Choy. Orange. Green. Red. Yellow. Chop. Chop. Chop little garden rainbow. I head to Jack's room and haul out the thirty litre bucket we store bulk rice in, I crack it open, cup it out, breathe in the scent of jasmine, then haul the bucket back.

I snip a bulb of garlic from the braids dangling off the coat rack, dice a few cloves with ginger and drop the little flavour bombs into the wok. The smell is outrageously enticing and people begin to appear.

Andrew starts dismantling the crockery towers, taking the racks to the ute to go back to the shed tomorrow. Lucy sets the kitchen table for seven. I now have bench space and some headspace, just in time to greet Maple and Leaf who have obviously remembered the 6pm dinner time I'd rattled off when welcoming them this morning.

Maple is your average Canadian guy, tall, healthy and good look-ing, Leaf is the opposite, her porcelain skin repeatedly interrupted by the sharpness of her bones.

"Hey guys, you've already met Andrew, but this is Lucy, Jack and Rosie," I say. I calculate in my head the kids have already lived with about sixty different backpackers, I wonder where these two will end up in their headspace.

The first dinner is always a bit of a feeling out, finding out the basics about their lives, why they're travelling, what they like and trying to make them feel at ease. It's a bit like dead-end speed dating because no matter how it goes, you know you're going to have to make it work over the next weeks and months – we've never kicked anyone out, when we take people on board, we take on the responsi-bility of making it work too.

"So, welcome! How long have you two been together?" I start with the usual question for couples. They'd applied together by email just four days ago and as we'd had another couple not show, and with the fiddly rosella flowers needing harvesting, we'd taken a punt and said yes.

"Well...we actually just met on the plane coming over five days ago," says Maple.

Andrew and I can't control our eyeballs and they rush to meet.

"Oh?" says Andrew. "We thought you were a couple." I know that is Andrew code for: *what the hell?*

So far, our worst experiences with Wwoofers have been when we've taken on singles or people who have just met. There was Griselda the passive-aggressive Londoner who would literally elbow

Andrew to get in a door before him, then plonk herself in my favourite chair helping herself to my stash of plonk and chocs. There was Hello Kitty, the Japanese baby doll who high-giggled and wide-eyed for a whole three weeks lifting not a finger except to depress the shutter on her camera. There was Madame Lafarge who did every-thing meanly and furiously...knitting, whinging, eating. And personal favourites, Gypsy the Irish lass and Elle the Eastender, recently bound mates who could make a multitude of fifty dollar notes disappear.

I pass the stir fry plates around.

"Thank you," says Leaf, and she begins to separate the vegetables from the rice. "Yes, we find out on the plane we want to do the same things, first do some Wwoofing and not spend any money, then to the Gold Coast where we can make money." I watch as she makes sure every grain is vegetable free before daintily scooping a few beads of rice at a time into her mouth.

"Sorry, you don't like stir-fry?" I ask. We try and get food prefer-ences and allergies out of the way on the first night so we can plan meals everyone will enjoy.

"I only eat brown and white food," she says. Say what?

"Like Nutella and ice cream?" Lucy's eyes sparkle at the thought of joining in a dinner like that.

"Yes," says Leaf.

"Like Vegemite and toast?" asks Jack.

"No, Vegemite is too black," says Leaf.

"Mashed potato and steak?" asks Andrew.

"Potato yes. Steak if it is brown."

I feel like I'm in an eating disorder Twilight Zone, my head clut-tering with images of Maltesters and cream, brown sugar and white sugar, pretzels and salt.

But dinner is just the entrée to her quirks. The following week when we run out of washing powder, I realise it's because she washes and irons every item of her clothing every day, whether she's worn them or not. She has bursts of frenzied smiling but takes to sleeping most of the day, and though I try to feed her up, she's basically too weak from the self-enforced Brown-White Diet to exist. Maple, who's gone from the guy in the seat next to her on the plane to her wannabee boyfriend and carer, apologises and tries to work doubly hard to make up for her. She raves she's the Paris Hilton of Canada, the daughter of a wealthy industrialist. She simpers and pouts and I begin to give the kids their dinner early so nothing rubs off.

Meanwhile, the Shetlands rub their butts against their corner fence post so hard, the wire slackens enough for the entire herd of goats to escape to all corners of the farm. After hours rounding them up, Andrew and I agree it's time for the Shetlands to find a new forever home. We spend the week preparing Lucy for their exit. There are tears, there are tantrums, but she can see a bigger pony on the horizon so it's finally agreed she'll ride Cindy into the ring, and tow Cody at her side, at Saturday's Gloucester Sales.

It takes two hours and two bales of hay to load them into the hired horse float, and the whole time my cheeks flame with the guilt and shame of selling them like objects, but I also feel relief we've finally made the call, that there'll be two less things to clutter things up, to clatter about.

Thunk. Thunk. They kick at the sides of the float. *Thank. Thank.* This is nearly over.

A few hours later, Maple and I are picking ruby-coloured rosellas on the hill when my mobile rings.

"Lucy did great," says Andrew. "She rode Cindy beautifully and Cody trotted perfectly alongside."

"Did you get an offer?"

"Yep."

"Great!"

"One offer...it was from the Doggers."

My heart sinks. The Doggers aren't a nice family with cute little pony-loving children. The Doggers aren't a nice family at all. There are three types of Doggers in Australia. There are Doggers who seek 'like-minded adult action' in public carparks. There are Doggers who use their canines to hunt and kill feral pigs. And there are the Doggers at the sale yards, who like the canners in the US, buy thousands of horses each year to slaughter for the dog food industry.

The Doggers...the thought of Cindy and Cody being decluttered of their heads, hearts and hooves, then processed and minced into tin cans...thunks me in the chest.

"Can you bring them home?"

"Already on our way," he reassures. And so am I, over to the feed shed to get them some welcome-back-I'm-sorry-they-only-wanted-you-for-dog-food, I-can't-believe-I-wanted-to-get-rid-of-you, let's-make-up hay.

I walk by the laundry door and hear the dryer going. Leaf is sitting on the floor watching it go around and round.

"Hi Leaf, the dryer is only for wet days," I say, perturbed at the energy waste, the world's biggest clothes line just metres away.

"I don't want any spots on my clothes," says Leaf. "If I hang them outside, they get spots. This place is so dirty."

Fragile, eating disorder, OCD, barking mad...that's how I'm going to end up if I put up with this any longer. I might not be able to declutter the paddocks, the lounge room or the kitchen, but I need to find the strength to declutter my life. I take a deep breath. A deeper one. Another one. I suck in oxygen like an Olympian at the end of a race.

"Leaf," I exhale calmly, "I get the sense you might like to head north to start your Gold Coast dreams. That's okay with us, I can get you to the train tomorrow."

"Thank you," she says, hugging me with the most strength she's used since she's been here, "That would be so great, but can we have a family dinner before I go?"

* * *

For a recipe for a brightly coloured and flavoured rosella cordial, see page 235.

FARM FENG SHUI

*I*t's bloody hard trying to cater for every personality, every taste, every species, every nuance; it's exhausting and blood-pressure raising trying to appeal to and appease while always feeling appraised. That's why we take an emergency two-night break to Sydney, surprising the kids with their first ice skating experience.

We're at the Macquarie Centre in Ryde, the kids entranced by things we don't have at the farm...things like steps, escalators and clean surfaces. They've finished their ice skating session and we're up in the food court. I'm marvelling at eating what we haven't had to grow or prepare, wondering where all this food comes from – and how the waste can be returned to the soil so future generations will get the nutrients they need – when Jack says innocently:

"We're so lucky we got into this food court without a ticket Mum! It's a really popular place for tourists."

I look around at all the Asian and South East Asian faces going about their grocery shopping, sipping their chai lattes, downing dumplings, kebabs and meat pies.

"Darling, these people live here, they're not tourists, they're Australians."

"Really? Well they look like the tourists who come to the farm," he says, and goes back to eating his fried rice.

I can see why he's confused, he's never been exposed to a capital city melting pot. Our area of rural Australia, being mostly white and indigenous, means his only exposure to Asian faces has been the growing number of visitors to the farm from Singapore and Hong Kong. The bulk of these tourists have been great, but some visits haven't gone swimmingly, so it prompts me on our return home to book into a government course on how best to provide for foreign tourists.

It's like being back in school again.

"Nothing scares the Chinese more," says the escapee from QANTAS stewarding, a prim male trainer in gold-buttoned vest, "than being in accommodation far from reception, and with a wide-open view. You see, they're used to skyscrapers close in, so it's best to plant bushes or put a fence up to make them feel more contained."

Strike one for us, we won't be blocking in the magnificent valley view.

"Saving face is a big thing for them," he continues, "don't give them a large serving of anything in case they don't like it and are embarrassed in front of their group – give them many small samples."

I'm already drained washing up bus group cups and plates, if I have to wash up another six cups per person so they can try all our teas, I'll go bonkers. Strike two for us.

"You need to be most careful about numbers," warns the trainer. "The number four means 'bad' or 'death', so never give a guest four gifts or they will take it as a threat or bad omen. Two gifts are okay, it signals balance and harmony."

I'm alright with that rule, we can't afford to give them four of anything anyway.

"Never have pears in the fruit bowl, the pear is a sign of separation."

"What if they're really, really awful and we want them to leave?" I ask. "Do you give them four pears?"

The trainer glares at me, horrified, but so am I, remembering the last lot who left dirty nappies on the picnic tables for Andrew – and chickens – to collect, filled our giant recycling bin with plastic water bottles in a record eighteen hours, and woke us at 11.45pm for a hairdryer for their two-year-old daughter's one square centimetre of wet hair. "It's still 34 degrees Celsius for goodness sake," Andrew had muttered, "why don't they just stand her outside for a sec or in front of the fan?"

I'm wondering if I should mention that incident to the trainer, but he hasn't recovered from pear shock and refuses to look at me. "You must, I insist, you must avoid the colour white at all costs, white is the colour of funerals and death."

Ah ha! So that explains why the visitor on our Meet the Bees experience didn't want to wear the white beekeeping suit and insisted he would be fine with a can of Aeroguard.

"Sir," I had said firmly to the customer who is always meant to be right – my inside head shaking backwards and forwards with disbelief, my outside head trying to stay perfectly still. "You are not going anywhere near my bees with insect killer. Do you understand? You need to put it back in your car." I cross my arms into an X and point at the can. Then jab my arms again, X. X. X! He nods, takes the can to car, and comes back with a DEET filled bottle of RID.

X. X. X!

Between my face, my arms and my stamping foot I finally get the message across and save the lives of about two hundred and fifty thousand bees, but set back Asian-Western relations to the Opium Wars.

Returning from the course I fill Andrew in on my new learnings about customer relations, and though we don't plan to meet all the criteria, we do decide to do a little Feng Shui out the front of the farm shop.

Feng (wind), Shui (water) is the ancient Chinese art of balancing energies in the environment to add to the wellbeing in your life...in our case, hopefully getting tourists to come into the farmgate shop rather than driving straight in and out of our very barren looking carpark. Unfortunately, though, because we've been living for the last seven years with a jade bush at the back door of our farmhouse rather than the front, money's been flowing out rather than flowing in, which is why I only have a budget of $90 for the new project. But low budgets are good for the environment, it forces us to get creative and recycle more.

The shipping container shop, once you come through it and out the other side, has a fabulous view of the farm and valley, but from the carpark it looks like an outback mining camp.

"We need something to draw people to it," I say.

"Not a pear tree obviously," says Andrew.

"We could definitely go a mulberry though, that'll give great shade in Summer too."

"How do we get all the Feng-Shui colours? Mass flower planting?"

"Too hard to maintain."

"How about we do stacks of painted tyres," bright sparks Andrew. "We've still got all the ones you used for the gourd planting, and we can get more from the mechanic, he has to pay to get rid of them otherwise."

The billion tyres produced every year across the planet create one of the biggest disposal headaches on earth. The burning of them creates highly toxic pollution and as more and more people drive cars, the problem simply accelerates. One day, Andrew and I would like to use them to build an eco-friendly Earthship house, but that's a project for another lifetime, today's project is Feng Shui and I need to stay on track.

Tyres aren't really safe to grow food in due to leeching of toxins, but we decide as planters for decorative shrubs and flowers, they'll be perfect. With the help of gorgeous British Wwoofers Ben and Charlotte, we soon have towers of tyres in bright blue, green, purple, pink, red, white and grey – symbolising knowledge, family, wealth, relationships, fame/reputation, career, children and helpful people. We angle them like a curving stream, creating flow where there was none.

It works like magic and more people begin to meander in, meaning we must get used to dealing with even more cultural nuances.

There are the special needs kids and special needs adults who come with their carers. Huge big smiles, wandering eyes, boisterous and shy, aggressive and amiable. Once you get used to their ticks, their Tourette's, their fetishes and furies – like pulling out all our plant sign markers and launching them at other plants, so the banana becomes a bunya nut, the cinnamon myrtle becomes cork, the pine nut becomes a pomegranate – they're a delight and honour to serve. Their heart-curved smiles leave us with warm and fuzzies for days.

I love too when families come through the door, the kids hopping from foot to foot in anticipation, and I ask them "have you ever been a farmgate shopkeeper before?" None have, so I invite them behind the counter and teach them how to use the register and get their own change, then send them on their way with an ice cream cone full of animal feed so they can make a farmyard friend fast. I love seeing their eyes go wide with delight when they munch on sweet stevia (*Stevia rebaudiana*) or I show them how to chew on toothache plant (*Spilanthes acmella*) to make their gums go tingly and numb, or when they discover native bees for the first time. I love that.

Then there are the people who try to con and cajole, "it's my daughter's birthday, can she please just pop in to see the animals? And can we take photos, we don't want to pay to come in though, it's her birthday you know. Her grandparents and cousins are coming to meet us too." These people get all offended and blustery when I say "no, sorry, we offer birthday parties or you can pay to come in." I want to tell them half the entry goes to insurance and the other half to the upkeep of the farm, and that we're basically volunteers. I want to ask them – when they leave in a huff – if their request normally gets them in free to Taronga Zoo and Disneyland?

There are the passive aggressives who bring the same out in me when they say nastily to our kids who are manning the front counter, "oh, so we can visit the shop, but we have to pay to go on a farm tour and meet the animals? Really? That seems a bit rich."

And I want to say, but hold it in, "you have the brochure in your hand which is how you found us, so I'm just wondering how you missed the bit about the entry fee which is written in **bold**. And is also written in **bold** outside the shop and again in **bold** right here at the counter. If I, a perfect stranger, asked you to miss a day of paid work so I could turn up to your house, be mean to your kids, wander your gardens, pat all your pets and spend three hours sucking information from your brain – and not offer you something for your time and effort – would that be rude or just **bold**?" I want to tell them our dream is to open the place for free so everyone can learn about solar ovens and bees and organics and permaculture but we just haven't been brave, smart or **bold** enough to work out how to do it yet...but if they'd like to give us some suggestions or become a benefactor we'd be all ears."

Then there are the delightful people whose eyes go misty as they soak in what nature has created, in plants and animals and views; but they're countered by the people disappointed by their giant expectations of our tiny budget farm. There are people who return again and again to enjoy the experience and others who steal again and again so they can take it with them...we 'lose' one hundred and twelve $10 metal tea strainers in two years, catch grannies with unpaid soaps in their handbags and beeswax candles melt away never having been lit.

I'm in the shop now, contemplating all this, when the glass door slides open and a lady tentatively puts her middle-aged head through the gap.

"Hi, welcome to the farm!" I say. "Feel free to have a browse or you can pay to come in and go on a tour."

She puts one foot inside.

I encourage her some more. "There are honey tasters there if you want to check out our latest harvest."

I can't tell if she is going to put weight onto her toes to come in, or back on her heel to exit. It's a see-sawing, flip flop foot, so I say, "there are some fresh laid eggs in the fridge too."

The weight of the eggs tip the scale! The second foot appears inside as does the rest of her body.

She seems to be looking at the soaps and balms and rows of honey, but her eyes keep flitting around. Looks at me, looks away. Looks at me. Scans outside.

"Are you on holidays?" I ask.

"Not really," she says. "I have a place up the road but I live in Sydney."

'Up the road' can mean anywhere from five hundred metres away to fifty kilometres, and I don't recognise her.

"Are you just browsing or did you want a tour? There are interesting plants and you can learn about native bees…"

"Well," she says, "actually…I just wanted to take a look at the place…it's just that my neighbours, lovely people, well, they swear blind you are nudists and that you're running a nudist colony here."

I nearly split my sides but hold it in because I don't want to risk showing any flesh.

This is the latest in a long line of things people have thought we are up to. Someone once asked if we were running a sect or a commune because of all the backpackers. And someone once asked if we were using the farm as a front to grow illegal drugs and someone once asked if we sieved our dam to get it so clear. But I have never been asked this before…though I do know the rumour's origin.

Seven years ago, we'd built a big earth mound in front of the house and yards to give guests and ourselves privacy and peace from the increasing traffic. One day, Dave, the local courier driver and expert deliverer of parceled-up gossip, dropped in with an explosive package.

"You got everyone talking with that wall," he said. "Some reckon it's a dam wall, some reckon you're building a bunker with a secret tunnel…but I've had a bit of fun and set 'em all straight – told 'em you're naturalists and you're running a nudie resort behind it." Ha, ha, ha! Ha, ha, ha! He belly laughs, his own package jingling.

Ha, ha, ha! Ha, ha, ha!

And here we are, seven years later, ha ha ha! Ha ha ha! With four well-publicised tourism awards, full disclosure on Facebook, Trip Advisor and Instagram, happy fully-clothed customers from around the world, and the rumour still rules.

"I'm sorry to disappoint you," I say, because for this customer, I definitely can't give her what she came for and I don't want to strip myself down any more.

* * *

FOR A MULBERRY CRUMBLE RECIPE YOU'LL CRUMBLE FOR, SEE PAGE 237.

DAFFY

*D*own in the valley, out of sight of the house, the cattle call to each other across the night. It's the restless lowing of twenty cows – sporadic and haunting – and moves Andrew to push his just-fried zucchini fritters aside and grab the keys to the quadbike.

"Want me to come?" asks Rosie.

"I'll call if I need you," he says, unsheathing the big Led Lenser torch from the recharger, pocketing his head torch too and heading out to the quadbike.

"I'm glad," she says to me matter-of-factly, "I don't want to find a sheep with no butt again." She puts her fork down, and I give her a gentle squeeze on the knee. Two weeks ago, we'd had our first wild dog attack in eight years. Dingo-crossbreeds are particularly vicious and when taking it upon herself to look for the sheep's butt, she'd found a lamb's head too. Both the butt of the mum and the body of the bub remain missing.

"It won't be dogs," I say to reassure her, "and the sheep and goats are in the extra-safe paddock." She takes that in and goes back to eating, until a few minutes later when the phone rings.

"Daffy's in the waterhole," he says, "need you, the kids, rope and hay."

Lucy's away so the three of us pile into the ute, stopping at the feed-shed for rope, and so Jack can help me load two bales of hay. You'd think dry grass would be light, but as we hoist them up they're heavier than tree stumps and our palms are pinched red by the taut baling twine.

Heavy cloud separates the moon from the earth, and even with high beams on, I struggle to stay on the track. We're headed half a kilometre from the house to where the creek winds its way through thirty-year old re-growth forest. Trunks of ironbark, tea tree, paper-bark and grey gum bollard up from the earth within an arm's reach of each other – their chunky bases and boom-gate arms creating an impenetrable barrier for the ute.

We get out of the car, I loop the coiled rope over my shoulder and Rosie torches for us as we navigate the dam, trees, ant mounds and clumps of blady grass. Under the torch light the blady grass is blunt red. When short and young it offers good fattening feed for cattle, but when it gets leggy, topping 200mm, it's like a blunt sword to the teeth of animals who eat it. Perhaps the grass is why Daffy ended up in the water – tempted instead by the succulent, fleshy reeds beckoning like lollipops from the waterway.

As we near the waterhole I feel the agitated vibrations of the herd rise through my feet and into my thighs.

"It's scary," says Rosie. I feel it too: the breaths of many in the dark-ness and the leathery sound of hides rubbing nervously against each other. The three of us cluster together like the cattle, safety in numbers, vibrating our own uneasiness back into the dark.

"Over here," calls Andrew, and I catch the glint of his head torch.

He's crouching at the biggest of the three waterholes, and following the beam of his torch, I see her. Her tawny head, the only visible thing above the opaque green of the water, circles slowly, ghoulishly, disembodied in the murk.

Normally the creek offers a shady haven of trickling water, moss and reeds. Sunlight filters through the canopy of trees, flecking the water with gold and lighting the banks a brilliant emerald colour. Along the creek are three waterholes, their depths unknown because we've never been able to hit bottom with any tree branch or pole. We've heard but never seen the platypus who live in the banks, and the kids used to test their balance skills across bridges of slippery, long-downed logs.

But it's not a place of play now. The long dry has cracked the linking creek bed, and the waterholes have dropped two metres from the banks, leaving a ledge exposed from which the cow has stumbled or slid in. Even with the two-metre water drop, her feet can't touch mud and it's like she's stewing in a bottomless, sheer-sided vat.

"What's the plan?" I ask Andrew.

"That's her baby there," he spotlights the fluffy white calf, standing at the outside of the herd fifty metres back. "We gotta get them all out of here if we're going to work safely though. Hay?"

"I'll get it," says Jack.

"Just a couple of handfuls mate, they'll follow you out, then toss it to the left so you can pull the wire across so they can't get back in. Make sure the calf goes with them. Then bring us some for her."

Jack dissolves into the night and I look back at the silently swimming cow. Daffy's part of the second herd of miniature Galloways we purchased to add to our fluffy crew. More timid than our gang, they normally baulk when we approach, not letting us touch them, but when I kneel at the side of the drop-off, her head comes directly to me. I hold onto a sapling to steady myself, reach out and stroke her forehead. I've never seen such knowing, such desperation in someone's eyes before, it's like she's screaming "help me!" but the only sound I can hear is the ripples of water swirling up from all the treading she's doing below.

"We're going to get you out of here," I tell her, tell myself.

Often a cow can be swum to a bank where she can get her footing, but that's not an option here. Nor is bringing in a tractor with chains to drag her out – a tractor can't get within three hundred metres of the place – and with her weight and the vertical height she'd have to be pulled, it would dislocate her hips and kill her anyway.

"Thanks mate," says Andrew, as Jack brings back the hay. "Hold onto Mum and hold it out to her." He does and she licks at it, curls a few strands into her mouth then swims back to the other side of the waterhole, then back to us for more, then back again.

Her calf calls through the forest and she lifts her chin and bellows back.

"How are we going to get her out Dad?" asks Jack, his voice trembling slightly.

"Don't know yet mate, that fallen tree is blocking her from swimming to the next section," he points to our right. "If we can just get rid of those branches she might be able to swim up and find something to get her feet on. Let's give it a go."

Carefully we cross between the waterholes to the other side. The bank is steep, the undergrowth scratchy and we pull ourselves up using the supple branches that hang overhead. We're about four metres above the water now, right where the obstructing tree has fallen in from. Andrew carefully feels his way down it and tries to stomp one of the branches off. His knee raises high and he drives it down, down, down; three times, eight times, ten times but it doesn't budge.

Daffy keeps swimming silently below.

"You'll be right girl," I call out, hoping to buoy her verbally even if I can't do it physically.

"This isn't going to work Dad, this isn't going to work," says Jack.

"I want to go back to the car," says a teary Rosie.

It's horrible to watch an animal fight for its life and to feel helpless to save it. It's worse when your own children are watching it with you.

"Take Rosie back to the car and ride the quad up for the chainsaw Jack," says Andrew, "Drive safe and slow, we need it – and you." With the kids gone I cross back to the other side of the creek, tempting Daffy with hay, trying to keep her strength, and my spirits, up.

When the first light drops of rain fall onto the dry leaves on the bank, it sounds just like dry leaves starting to burn. It's not pitter patter, but a hiss and a crackle. Hisssss. Then it builds, and droplets of bleak sputter from the sky. The sweet sickly smell of rotting reeds and mud fills my nose, and the sickly sight of Daffy's head getting lower in the water bleaks me further.

Andrew paces the bank, slowing only when the sound of the returning quadbike rises above the rain.

Just as parents shouldn't let their kids jump on trampolines with fishing knives, or run with scissors, a parent should probably not encourage their children to run with chainsaws, but here I am yelling "Go Jack, go. Faster. Come on!"

Andrew starts it up and the urgent, ripping NINNNNG – the sound and smell of dieselly metal coming to the rescue – makes my hope surge.

Except he can only reach one of the branches. He twists and contorts, leans and stretches, but can't get the right angle to chainsaw the second branch free. He sets the chainsaw to idle and puts his face in his hands.

"Mum," Jack whispers like he's telling a secret, "I don't want to watch her go under."

I follow his eyes to hers. She's looking colder, weaker, lower.

"She's not going to go under mate, we're going to get her out," I say, but my stomach churns and I realise this might turn into a great life lesson for the kids – about never giving up, about helping creatures, about working as a team...it might become 'the night of the legendary waterhole cow rescue', a cherished family story told by him to his children and them to theirs. Or, it might just become a nightmare he'll carry forever. My brain starts whirring, we need a human chain of helpers, a helicopter rescue, Santa and his sleigh, James Bond and a hovercraft...

"Let's get the pedal boat!" I yell.

"I feel sick Mum," says Jack.

"We can't give up mate, we have to try until we either get her out or she–"

"Come on, let's go mate," says Andrew, tousling Jack's hair, laying big hand on little shoulder.

We jog back to where the ute is parked near the dam. Rosie's still waiting, cocooned on the backseat. Her eyes widen when she sees us.

"We need you gorgeous!" I say.

We'd originally bought the pedal boats to encourage guests to have fun on the dam and explore with their kids. But out of hundreds of parents, just a few have bothered to voyage the lilac-lilied waters with their kids.

White and blue, the size of a double bed, the pedal boats weigh more than a king-sized waterbed. We hoist the front of the boat onto the quadbike's back rack, and with Jack driving the slowest he ever has, Rosie, Andrew and I follow like bridesmaids, holding the back of the boat off the ground. It's like trying to carry the Titanic. Every few metres we need to stop, reposition, give our muscles a rest, but there's some superhuman adrenalin at play and we get a few hundred metres before the trees block the quadbike's way.

"Put it on its side," says Andrew, and we hoist it vertically, then thin it through the trees, shoving and pushing and not caring if we wreck the rudder or ourselves in the process. Finally, we smoosh it through the reeds and onto the water. Andrew gets aboard.

There's no outboard motor, just his two feet on tiny blue pedals.

Pedal, pedal, pedal.

Pedal, pedal, pedal.

It's like executing a rescue with a blimp rather than chopper, a water pistol rather than a fire engine… "we'll get there soon ma'am… we're coming,"

Pedal, pedal, pedal.

Pedal, pedal, pedal.

What Andrew does next is something you should never try at home, though here we are, at home, doing it: he stands up in the rocking boat and pulls the cord on the chainsaw.

NINNNNG, NINNNNG. Wobble, wobble.

NINNNNG, NINNNNG. Wobble, wobble.

The chainsaw, blunting fast, sends sparks and shards flying, but eventually we hear the crack, he throttles the chainsaw back, and the branch falls away. As he nearly does too.

Wobble. Wobble.

Pedal, pedal, pedal.

Pedal, pedal, pedal.

"Can you tie a rope around her? Get her to swim alongside?" I call out, "or how about we put hay in the boat and we get her to follow you?"

Pedal, pedal, pedal.

Pedal, pedal, pedal.

He comes close to shore and Rosie throws him the rope, Jack leaning out with the hay.

Daffy's calf calls for her again, but mumma cow's reply barely registers, it's a sodden, black-wrapped moo, so hoarse and sad it's like a giving up goodbye. Her distress is so distressing my neck locks.

"Hurry Andrew, hurry!"

Pedal, pedal, pedal.

Pedal, pedal, pedal.

He grabs a floating stick, pulls up beside her and jabs, trying to prod her to swim to the new opening. The kids and I stand on the bank, armfuls of hay splayed out temptingly, calling.

"Come on Daffy, you can do it! Your baby needs you. Come on Daffy, it's so yummy. Come on, you're getting cold and it's warmer out here."

And Jack. "Don't die Daffy!"

She oozes slowly to the gap, slips through and buoys over to the bank where our outstretched, hay-laden arms try to get her to stretch some more. She bobs around the new space before going to head back through the gap.

"Stop her Andrew!!!!" I yell.

Pedal, pedal, pedal.

Pedal, pedal, pedal.

Andrew's like the goalie in an Olympic water polo match, pumping with adrenalin and blocking Daffy's own-goal at the last minute.

"Daffy, Daffy!" calls Jack. She floats toward him but, like a furry crocodile, it's now just her filmy eyes and the top of her snout above water. Jack runs around another curve and starts kicking at another downed branch that gates her in.

Kick. Kick. Kick.

Half of it comes apart like cork, but the other half still blocks Daffy from testing another get out of jail escape route.

"Throw me the rope Dad, throw me the rope!" Jack yells.

Andrew stands up, wobble, wobble, wobble, wobble, and throws it ashore. Jack catches it, loops it over the blocking branch, tugging stronger than his young boy arms are capable of, till it finally breaks free and swings back toward him.

I don't know what more we can do. There are so many underwater obstacles seeming to block her, so many hours, so much exhaustion. The cracked mud at the edges suck at our boots, making it hard to move quickly, or even slowly for that matter, I can only imagine how dragged down she feels too. But no one's given up yet.

"Come on Daffy!" I will her with words, with fear; not wanting to regret, not wanting to see her sink, not wanting her or us to lose this battle.

Everything's salvageable…until it isn't. Any moment can turn to gold or mud.

Daffy sludges around the bend, leans her chest into the bank, rests her chin. Rosie doles out hay like crumbs for Hansel and Gretel.

"Get the herd in," says Andrew.

Pedal, pedal, pedal.

Pedal, pedal, pedal.

"Get the calf."

I sprint to the wire and unhook it. The herd, eyes reflecting in the failing torchlight, mumble and shuffle, not knowing if they're being invited to a wake or an awakening. The little calf moos, a plaintiff call for milk and mum.

I see Jack's face before I see Daffy's. It's the face of a lotto player with five of the right numbers, one more ball to drop. It's the face of a basketballer watching the potential buzzer beater fly through the air. It's the face of a father waiting for his new born baby to make a noise. Any noise.

The noise I hear is squishy.

Daffy has her forelegs in the mud, pulling herself up on a submerged ledge. Millimetre by millimetre she follows Rosie's trail of hay, until her rear legs, buckling from the cold water, exertion and stress, begin to heave out too. The furry crocodile rises from the mud transforming into chocolate milk.

There are reunions all round. The calf rushes in, suckling even as Daffy shakes from exhaustion and the hours in the cold water. We humans hug and jump and yahoo, I feel so light, like I'm wafting a few centimetres above the ground.

"We did it, we did it!" says Jack, and it dawns on me we did.

Leaving the pedal boat floating in the waterhole, we slow walk behind the cows until we can stretch the wire to keep them out of the danger zone. At the ute, I take my boots off, upend them, a stream of water and adrenalin pours out, and a stream of happiness pours in. I rifle through the CDs until I find the one I want. Winding the windows down, Andrew doubling Jack on the quad alongside, we motor home to Sister Sledge's 'We Are Family'.

Tonight does get to go down as the legendary waterhole cow rescue. It goes down as a night the kids learned legends only get up, when you don't give up. And it goes down as a night when a rescue brought family – bovine and human – even closer together.

* * *

FOR A RECIPE FOR THE FRITTERS WE LEFT AT THE TABLE THAT NIGHT TO rescue Daffy, see page 239.

PIXELLATING

I'm short sighted at the best of times, but even with my specs on, things are beginning to blur. There's the combined haze of farmstay guests, backpackers, day visitors, farmers market customers, locals, shoppers and bus tour groups. I calculate I've had multiple-sentenced conversations with more than thirty thousand people in the last few years alone. My noodle is a brimming alphabet soup of the Latin and common names of plants, so full and slurpy there's no room to commit human names to memory.

Adding to the blur is the interaction with multiple animal generations: is that cow Mudcake or Cupcake or CocoPops? Is that goat Bubblegum, Vanilla or Soloette? Is that lamb Blackjack or Jedi or Skywalker? Is that rooster Rocky, Rocky Two, Rocky Three or Cliffhanger? Actually, it's not hard to recognise Cliffhanger, he's the one who attacks your throat with his razor-edged spurs then goes for your eyeballs with his beak. I've probably patted, chatted and connected with more than five hundred animals – not including bees – who would rocket that number into the millions.

When I look in the mirror I fear I'm beginning to blur too. Having to engage with and adapt to so many beings – from so many different backgrounds, with so many different needs – is pixelating me. I feel for the bartenders, the counsellors, the teachers, the checkout operators, the nurses, anyone on the front line. I feel for me right now too, because my work is my home and my home is my work and I can't get away. That's why today I'm camouflaging behind the sugarcane, planting sage, and waiting for the farmstay guests to find me, rather than seeking them out.

It's so beautiful out here, so peaceful. Rose petals syrup the air, lemon-scented tea tree drizzles zest, curry leaves surprise with sweetness and spice, and the sun and breeze grapple with each other in a warm, ruffling embrace. The freshly planted seedlings of Salvia officinalis, pungent with camphor and no hint of the lemon they spark on the tastebuds, poke soft green and silvery above the dirt. I pat the soil around them, wish them well, and water them in.

"Farmer Anna? Farmer Anna?" Here they come, little blondies, full of spark and fun. It's not really them I'm avoiding, but their parents. Andrew's been hosting them the last few days while I've been at markets, and now it's their last morning, it's my turn. When I asked how the parents were he'd said, "I haven't seen much of them, they just send the kids out," which is code for he's not impressed but he's not going to build up the tension for me by foreshadowing the interaction.

"Hey kiddos," I say, the excited expectation on their faces making me feel guilty for sugar-caning, but before I can say, "stay on the path", mini-surfer-dude, the human grasshopper, bounces onto the baby sage. His little sister follows, foot-printing the survivors into the dirt.

"Can we feed the alpacas before we go? Can we, can we?" he asks, hopping from foot to foot on my morning's work.

It's times like these the blurriness helps – it's just another event, not a penultimate event. I don't really get upset at all, just silently wince and feel my insides pixelate a little more. He's so cute and keen, and I comfort myself with platitudes that *it wasn't on purpose*, and *it doesn't matter* – and at least they weren't like the last guests who pulled the shower rail out of the wall, put their foot through the flyscreen, left boot prints on the bedspread and went out one day for three hours…without taking their 2 and 4-year-old with them.

"For sure," I say, dusting off soil and flecks of *I wish I could just finish what I am doing without being interrupted*, while wondering how Andrew will react when he sees the feed, repair and seedling restocking bill this month. School holidays are when the cash flows in from farmstay and day visitors, but it's like a tsunami: the big dollar wave comes in, and just when you think you're safe, rushes back out taking everything with it. Then it's a long three months till the cresting wave returns.

But you can get ground down or you can grin up, so it doesn't take much to re-energise me – kids do that for you. Arms outstretched, I pretend to glide on the breeze, the blondies and I laughing as we soar along winding paths to the animal yards where we lean on the splintery wood, catching our breath.

Alpacas are the push-me-pull-me of hilarious animal jokes: a combo giraffe-bunny-cat-goat-camel-koala-sheep-Pokémon fantasy creature that you want to cuddle desperately – until the moment it spits vile mucousy grass into your adoring open mouth.

Bravado is the leader of the herd: heroic, broad-chested and calm, he's the trusted sentry at the castle gate, and the opposite of teenage rebel Cloud, who, if he were human, would be drinking JD and Coke and doing donuts in the carpark at midnight. The women of the herd are inquisitive and coy. Spring hides behind her mother Jarra's silken white skirts, just as Jarra still hides behind her mother Cienta's. All three of them step deferentially around the chickens, and fuss maternally over baby lambs and goats. Stud male Patch, ensconced in his lustrous locks in the neighbouring paddock, trots and prances like a rich playboy on the deck of a superyacht.

"See how these white ones have a fleece like dreadlocks?" The kids nod. "That's because they're a special type of alpaca called Suri. Long strands, really silky, really shiny – they're not woolly like him." I point to Jasper, Jasper of the huge chocolate melt eyes, cocoa fleece and teddy bear huggability; Jasper of the Huacaya (pronounced Wah-KI-yah) type of alpaca known for its dense, water-repellent fleece; Jasper who will sit down amongst children high with ADD, let autistic adults rock alongside him and will ever-so-sweetly walk up to prams, gently anointing human babies with the softest of nuzzles.

"Let's get them an extra yummy breakfast," I say and the kids bounce after me like roos.

The feed shed is in the first bay of the fifty-year-old stables. Back in the day it would have made the ultimate equestrian statement with its waist-high walls of white-painted brick, topped by rough-hewn, coffee-coloured timbers climbing in fat slats to the corrugated iron roof. The stables are entered via heavy swing doors, the timber and metal framing sturdy enough to keep stallions and mares apart. But time has leaned in on them and the doors no longer swing easily, cobwebs hang like chandeliers and there's always a fine shimmer of dust in the air. The smell though is rich: apple cider vinegar, molasses soaked barley, and the sweet clip of lucerne and oats. It's a happy, nourishing place, and I look kindly at the cobwebs and the skittering starburst babies nearby, thinking of spider Charlotte and the stories of E.B. White read to me as a child.

I line up six empty ice cream buckets on the floor. For each blue bucket, I hand the kids a scoop of oats, bran, lucerne and barley; then a sprinkle of yellow sulphur powder to repel parasites, a teaspoon of mineral rich seaweed meal, a vital puff of selenium to make up for its deficiency in our soil, and a teaspoon of dolomite, a powdered-rock powerhouse of calcium and magnesium. I wet the mix with apple cider vinegar and olive oil, and the kids enthusiastically stir the mix with knobby twigs.

I force open the stable door with my hip, and we head back to the yards, each of us carrying two brimming buckets. There's a man already there, fair-haired and barrel broad, and leaning on the fence so heavily, I add another job to the end-of-school-holidays list: re-concrete post.

"Is that your Dad?" I ask surfer boy. He nods.

"Hi there, I'm Anna, Andrew's wife. Hope you've had a good stay."

Grunt.

"Dad! Dad! Look what we made!"

Grunt.

I take my lead from the kids and decide not to react to the grunts, maybe he's having a bad morning, maybe he doesn't want to go back to his city life, maybe this is he and his wife's last-chance pre-divorce holiday.

I slide five of the feed buckets under the fence, about two metres apart, to ensure an alpaca-phlegm war doesn't erupt over breakfast. The surfer siblings lift the last bucket together, reaching on their tippy-toes to offer it to Jasper.

"Your kids are great with the animals," I say to the Dad. "You like animals too?"

He scans me, gumboot to ponytail, and replies, "I like to shoot them."

My brain blurs. My tongue blurs.

He cocks his finger at Jasper, depresses the trigger finger.

The line blurs.

Is that his sense of humour? Or is he serious? And how should I react? Who am I in this moment? Right here, right now: am I the hotelier, the farmer, the animal lover, the stranger, the mother, the stooge? Do I take it seriously or laugh it off? How do I respond?

Lately I've become a human walk-in wardrobe, trying on a different personality outfit for each unexpected occasion. With every new person I meet I wonder what's the appropriate fashion here? What's the mood? How do I slip into the same uniform as everyone else? How do I walk this runway or get a seat, not in the front row, just up the back in the shadows, not seen but there? Not always having to engage, but happy to applaud and wish others well?

The me in me blurs, I go to take on his wardrobe. "You'd like this then," I say. "There was an alpaca association meeting a while back and the guy who's place it was at serves everyone lunch, everyone's loving it until the moment he tells them the sausage sandwiches were made from one of his alpacas."

He grunts with a bit of enthusiasm and I go to tell him another alpaca misfortune story but can't go through with it. I decide to strip, a back to basics costume change.

"Would you like a herbal tea before you check out?" I offer. He looks at me like I'm asking him to join me in swallowing Ten-Eighty fox bait. I look back, wanting to share through my eyes and heart what I'm thinking...that when we sip on tea we sip on eons of microbes, the invisible wriggles of worms, last season's fertiliser shovelled by hand and wheeled by barrow to the compost heap, a year's dew and cleansing rains. That the leaves have been dried by winds that have circled the earth since the beginning of time; winds that have filled the lungs of mammoths and sabre tooths, dictators and doves. I want to tell him that tiny insect tongues have kissed generations of this tea's flower forebears and the tiny feet of lady beetles have trodden the leaves on their path to beetle-dom. The boiled rainwater, absorbing all this wonder, is absorbed through our blood and gives us life, the nitrogen eventually returning to the earth to become tea for another, maybe centuries from now.

That's who I am and what I feel, and I think the brew would help him, but that's not going to be his cup of tea. I blur myself again and say: "if you don't like it, you can just tip it out."

Pixel by pixel, I feel a bit of me tipping out too.

* * *

How about you make yourself a cup of herbal tea while you read the next chapter, or head to page 241 to discover how to make the most of sage.

NOT SO SHEEPISH

*T*his week I'm learning to say 'no'.

I look into the polished aluminium reflectors of the solar oven and practise forming the word with my lips. I want it to become an indelible mould for things to come. "No. Noooo. NO." I try and say it without emotion, without malice, without heat. Just an emotionless, easy, no-guilt 'no'.

Not being able to say 'no', to have it quickly and firmly roll off my tongue, has been making our lives ever more complicated. It's one of the shortest and simplest words in the English language, but for people-pleasers like me, the softer and longer 'yes' is much more automatic to pronounce.

My latest 'yes' is making me regret it right now. It's a demented two-month old lamb we took in for a local couple. Most lambs have a bucolic baa, a sweet, milky tinkle, but this one is Fran Drescher on a megaphone – a surround sound chainsaw of vocal chords capable of shredding any eardrum within a kilometre.

It doesn't 'baa', but "BWAHHHHRRR!" All our sheep, with disgusted *whose idea was that?* looks on their faces, have shuffled to the furthest corner of the farm to get away from it.

If I put it down, it "BWAHHHHRRRS!" If I go inside for a second it

"BWAHHHHRRRS!" If I don't give it a bottle every twenty minutes, it "BWAHHHHRRRS!"

The demented chainsaw echoes around the valley. It's been going on for hours, and night is coming.

"Are you going to do something about that sound?" asks a guest politely, "my husband really needs his sleep tonight."

"I'll sleep in the trailer with it Mum," offers Jack, he's our go-to kid for night time lamb care. Even as a four-year old he would take control of any orphan, sleeping beside them on the laundry floor, giving them warmth, a heartbeat and bottles throughout the night, until they meconium'd all over his angelic sleeping face.

The trailer is a multi-purpose rectangle on wheels that gets towed behind the quadbike. Sometimes it has laughing guests on board, or spiky hay or dull blades or oodles of tools. Tonight, we fit it with a double foam mattress and a plastic sheet. While Jack gets his sleeping bag, I get the pillows and Andrew warms another bottle for the woolly megaphone.

Finally, there's silence as it suck, suck, sucks, but as soon as it takes the last drop it lets rip again:

"BWAHHHHRRR! BWAHHHHRRR! BWAHHHHRRR! BWAHH-HHRRR! BWAHHHHRRR! BWAHHHHRRR! BWAHHHHRRR! BWAHHHHRRR! BWAHHHHRRR! BWAHHHHRRR! BWAHH-HHRRR! BWAHHHHRRR! BWAHHHHRRR! BWAHHHHRRR! BWAHHHHRRR!"

The noise is so grating, so loud, so intense, my whole body vibrates. I feel the impending angst of sleep deprivation, guest complaints, and the black hole of knowing it's not going to end. I feel like a mother with a devil newborn.

"Why did they ask us to take it when they knew how bad it was?" I ask Andrew. "Who does that?"

We take turns stroking it, cooing to it, and re-fill the bottle with water so it can suck without getting tummy pains from too much feed. Just when we think it's settled, it lets rip again. One hundred lamb toenails scratch down a blackboard. BWAHHHHRRR!

8 pm. 9 pm. 10 pm.

"What should we do? What should we do?" I ask in a gnashing teeth sort of way. Do we take it for a ride in the car like we did when the kids couldn't sleep? Or put it in a pram and bounce it down a bumpy path? Or dose it with peppermint Mylanta or a syringe of Phenergan? Maybe it needs counselling, or restraints and a padded cell. Maybe I do too I'm so jittery.

"I'm going to take it back," Andrew says. "It shouldn't be our problem."

"At 10.30 pm at night? Really?" I ask, shocked at the simplicity of the solution but feeling awkward about letting someone down, the possible social fallout of my saying 'yes' then 'no', the awkward moment when we wake them from their sleep to deposit the wool.

"Yep," he says, steel in his voice and eyes. "Oath!"

Now the decision's been made, he doesn't muck around, grabbing the lamb, a cardboard box – and Jack to keep the lid on it. They pile into the ute and drive off, the intensity of the BWAHHHHRRRs accelerating but getting more distant every second. I don't think I've ever felt such enormous relief, it's like I haven't inhaled or exhaled for hours, and suddenly I can breathe again.

I toss up between a chamomile tea or a sauv blanc. I down two of each and go again. About 11.30 pm I hear the tyres roll in on the gravel.

"Oh-my-god Mum," says Jack, rushing into the house, "when we got there, it smashed straight through their screen door and went and lay down in front of their fireplace."

"You're kidding?"

"They were trying to get rid of it because their neighbours complained about the noise."

"Serious?"

Andrew nods and I feel my anger rising at being put in the position we were, and anger at myself for agreeing to it.

"Thanks you two, and just so you know, I'm going to be working on my no's from now on."

"You're actually pretty good at saying no to me," says Jack.

"Well I need to work on it for other people," I say.

And the next day I start.

"No, stop looking at me like that, you can't have another bale of hay – there's perfectly good grass out there."

"No, I'm sorry, it's only 9.30 am and check-in isn't until 2 pm, the previous guests haven't even left your cottage yet. I understand, but maybe drive via the lake and see if you can spot dolphins, then grab some lunch and we'll see you this afternoon."

"No, even though he has his own Facebook page, our petting zoo can't take on your completely blind, but fully entire billy goat."

By the end of the day I'm covered in compost, comfrey stains – but also glory – having successfully formulated the word "no" on three occasions. I'm even feeling confident enough to take on the answering machine.

You'd think it would be chirp, rustle, mooooo – but the main sound on the farm in the afternoon is blip, blip, blip. The world's most annoying answering machine blips with my mind, the only way to make it stop is to pick up pen and paper and press play.

Play? Hardly!

"Help! Bees are swarming inside our wall, you need to come. You do it free of course? We're only an hour away. Call us back now!"

"We're past guests, heading north on hols, you don't mind if we just pop in tomorrow for a few hours – the kids really want to see the animals again."

"We are from Italia. We backpacka, we luvva to cooka. You needa help? We needa pay."

"I met you at a farmers' market a while ago, can you tell me the recipe you use to make your lip balm?"

"You're kidding me. You don't answer your phone? What a joke!"

"Heard you folks got a pipe layer, can I borrow it?"

"We're booked into the farmstay tonight, we know your website says strictly no dogs because of the farm animals, but they're just two little Dobermans, we're leaving soon so call back if we can't or else we'll bring them as we're leaving Sydney soon."

"I ordered some honey from you last week and it arrived. It

arrived all over the post office. I am very, very upset. So is the postmaster. I expect a call back today."

"I'm calling about advertising in our school holiday guide..."

"You do the Manuka? You do China?"

"Would it be possible for Miss Teen Universe Australia to have some photos taken with your animals?"

Say what?

It's time for answering machine message triage. Message triage is a modern-day farm skill right up there with fixing fences, removing testicles, making compost heaps and saying 'no'.

To pass triage I need to ignore the walking wounded (advertisers, freebie hunters, abusers); immediately treat the life-threatening (guests with demands that can't be met, unhappy customers); calmly respond to less urgent calls where we may need to refer them on to other specialists (example any commercial beekeeper crazy enough to travel two hours return plus spend another four hours trying to extract bees of unknown pedigree from a wall cavity...for free).

Hmmm...but what to do about Miss Teen Universe? There's always someone who arrives in the triage department who's difficult to assess. Is this a yes, a no or a maybe? Might need a specialist.

"Andrew?"

Three days later Miss Teen Universe arrives, six feet tall and clad in gorgeous Boho. Mikey the cross-bred Dorset is the first prop. With his lanolin locks and fierce curled horns, he looks like a stud ram – but he's really a wether who thinks he's a dog. Mikey walks straight up to her for attention. She recoils.

"He'll wag his tail if you pat him," Andrew says. But her perfectly manicured hands are reticent to reach out.

"Never work with animals," mutters the photographer.

It's hard not to relax around Mikey though, and eventually the shutter captures the moment. The sheep owns the shot, absolutely nails it. Eyes locked on the camera, he's strong and sexy, and, based on this pic, if it was a judgement call for Next Top Model, he'd go through to the next round.

Soda the Pony is the next to star. He's hormonal today and is blaming it on his Cushing's Disease. The benign growth of his pituitary gland means his cortisol is way off the scale. It gives him fast growing, thick and wavy Beyoncé hair – but in his case, it just makes him look shaggy and sweaty, more like Mick Jagger at the end of a set. The raging hormones also cause hoof pain and laminitis, so a bit like a rock star, he gets drugs, in this case, dopamine. The disease makes him a great prop because he can barely walk, but the medication cost makes him the most expensive equine in all the lands. But what are we going to do?

"Finished shooting the cute pony?" I ask.

The photographer nods. It's time for the final outfit.

I don't notice fashion, I don't notice labels, I don't notice trends – but I do notice white. Except for my beekeeping suit I haven't worn white since I was in my twenties, kids and farming will do that to you. It's just not worth the stain risk. But I'm noticing white now because Miss Teen Universe Australia has changed into a white ball gown.

Not just any white ball gown, but THE one she will wear at the international pageant in a few weeks' time.

I hold my breath.

All I can see is danger! It's a war on white. Miss Teen Universe Australia is surrounded by weapons of mass excretion.

A mine-field laid by Ovis aries aka: sheep.

Goat grenades.

Cow pat cluster bombs.

Horse manure missiles.

Chicken droppings. Dropping.

Alpaca ambushes, firing from both ends.

And then I see the booby traps! The barbs and spikes of barberry and finger limes. Razor wire rose bushes. She swishes her way up the catwalk path. I imagine the tiny little pitchforks of the plant known as Farmer's Friend – because they stick by and on you till you pull them out – peppering her gown. I imagine the juice of a black mulberry being jettisoned overhead by a crow. I imagine her falling into a compost trench stew of sawdust, skeletons and last night's stir fry…

that my saying yes to her coming will end in her downfall in a dress that no amount of Omo or bleach could resurrect for the pageant.

Every white cell in my body wants to scream 'No, stop! No! Don't go out there,' but I just hold my breath.

Her crown, a good foot tall, glints in the sun. Her smile, a good foot tall, glints in the sun too. She hitches up her dress, walks through no-mans-land and unfurls her pristine petticoats.

She poses, I pray.

The photographer snaps some shots and while he's changing his battery, she waves to our younger daughter. In the kindest of voices she offers, "would you like a photo with me? You can wear the crown." Now Rosie's smile is a glowing, good foot tall too.

The shoot finishes, so does the stain danger.

Nothing. Goes. Wrong.

The shot of Miss Teen Universe and Rosie is magic, she is magic, and I realise that sometimes it's not a strict 'no' that's required, because good stuff often happens when you remain open to – but not a slave to – a 'yes'.

* * *

AUDREY HEPBURN SAID, "TO HAVE BEAUTIFUL LIPS, SPEAK KIND WORDS," but if they're dry, you'll need a great lip balm. Recipes are on page 223.

SOUL

*T*he problem with the simple life is it's false advertising.

Why? Because there are three perceived benefits of 'living the good life'. They are as follows.

One: you won't need any money because you'll be growing, making or swapping for everything you need.

Two: you'll learn how to make your own alcohol and be able to swig at pleasure.

Three: you'll feel like you're always on holidays so will never need to take one.

The reality though is this.

One: nothing these days is free, so you will never stop spending money, until of course you go broke.

Two: you need to make your own alcohol because you can no longer afford to buy it, but it will taste like Metho.

Three: you won't get time to drink it or go on holidays anyway because you'll be too busy doing all the above.

So, when one is trapped in a cycle, one cycles like mad and that's why I'm madly trying to come up with another product that will simplify our balance sheet...and then, hopefully, our lives.

I've settled my hopes on beetroot.

Beetroot is bulbous, bright and never fussed by neglect. It's a stripper, the leaves dance on the tastebuds in salads with fetta and walnuts. The beet can be roasted, relished or rawed, chocolate brownied, souped and juiced. It's a dye. And the condition it causes – Beeturia – can make you think you're dying. It's responsible for turning the urine and faeces of approximately 1/5 of the population – especially those with low stomach acid like self – the colour of a blood orange, and sending many to hospital thinking they're haemorrhaging. It can be a fun trick to play on people but the local emergency department tires of it quickly.

I think a lot about beetroot. We've had a local chef bottle up our first big harvest as relish, and now I'm experimenting making alcohol from it.

"Really? Beetroot wine?" Andrew, the non drinker in the partnership had queried.

"It would be so cool! It would be unique! We could get a liquor licence. I mean, look at how many people go to wineries, not everyone wants honey, honey! And beetroot – it's so good for you, it would be like a health drink with benefits! Seriously, this could be it!"

So now, in addition to backpackers reclining around the house, beetroot does too. It reclines in its 25 litre carboys – bulging glass barrels the size of a small child – and converses in a meditative blub-blub-blub as it metamorphoses into vegetable moonshine. There's a blub-blub-blub from our bedroom closet, a blub-blub-blub from Lucy's wardrobe and an answering call from the corner of Jack's.

On a faded page of an old homesteading book, I'd read that if sunlight sights the liquid, the beetroot wine will lose its burgundy richness, turning a festy rust colour. So I've wrapped each of the huge glass vessels in towels to block the light. With their constant chatter and cute towel robes it's like a house full of little stooped E.T.s. "Phone home. Blub, blub. Phone home."

Half empty bottles of cheap vodka are scattered about too. Not only does the stuff sterilise stress for some, it tops up the valves of the carboys, blocking unwanted air, yeasts and gnats from spoiling the potion. It also means the little bugs that do fall in die pickled. And stay so.

Yes, there's a lot wrong with storing bucket loads of alcohol in the kids' rooms – but I ignore the misgivings because they're still at the palate age where bubbling beetroot, cloves and lemon peel are culinary hell. Plus, they're going to be away from temptation for a while because we're going on our first major holiday in years!

The workload and the care load the farm requires skews and compresses the years. Any three days off in a row, in any given year, has been classified as our annual holiday. It's weird that guests are on holidays all the time with us, yet we never are. For the first six years, everything at the farm was such a joy, such a revelation, it felt like we were on permanent time off, but the last twelve months have been especially difficult, and we need to get away, get the kids out of the workplace and to a place where life won't be a dawn till dusk chore.

"The Gold Coast!" We all agree, a place where irresponsibility rules, where five days of freewheeling, fun parks and fast food sound like just the ticket we need.

The farm provides such a confusing contortion of life, work, love, responsibility, hobbies, freedom, gratefulness and neediness, it makes it almost unbearable to leave it, and unbearable to stay. But we force ourselves to wave goodbye, leaving a crack team of four Wwoofers in charge, and drive ourselves north. The kids complain as we stop at every farmgate shop and farm attraction along the way – looking for inspiration, seeing what we've missed – but their eyes reveal the pain was worth it when fourteen hours later we step into our room on the 56th floor of the Soul Apartments on the Gold Coast.

"Do you think we'll see a Hummer?" asks Jack.

"Do you think there'll be a clothes shop Mum?" asks Lucy

"Do you think it will fall over?" asks Rosie.

I feel giddy looking down at the veins of traffic with their constant red-lighted arteries. I feel in awe looking out to the ocean, and imagining the wind rocketing in from the vast Pacific. It's like being enclosed in a snow dome of luxury where the hardest thing we must decide is where to eat and what theme park to go to first.

It's a revelation how tantalising all the food choices are, especially when you haven't had to personally shovel the manure to grow it. It's amazing to still be in bed at 7am, without a guest or backpacker knocking at the door. It's fun to see the kids laughing, the only exertion we're asking of them to stay in line for the next ride.

But each night back at Soul I sway at the height of the building, search the spread of Surfers Paradise, unable to find its depth. I'm at a loss as to what to do with the leftover milk in the kids' bowls, the orange peels, the cores. There's no chooks or compost within a concrete mile and I can't believe how much I care. I wonder how tower-dwellers exist and connect, when their feet can't touch the ground; when nothing needs attention but oneself. It's funny they called it Soul, it's funny they called it Paradise, because it doesn't seem to have one and it doesn't seem to be one – but hell, it has been fun!

"Relaxed?" I ask Andrew as we start the drive home.

"Yep, just a little worried about the animals," he admits.

"You?"

"It was great not being on someone else's timetable."

"Got that right!" He glances over at me. "You look a little worried though."

Worried we can't do this with the kids more often.

Worried about the bills.

Worried the beetroot wine's exploded all over the house.

"Nope, all good!"

Why have we made the simple life so complicated?

* * *

BEETROOT WINE IS A BIT COMPLICATED TO MAKE, BUT THE RECIPE FOR A beetroot relish is on page 243.

#NOFILTER

*B*ackpacker Bambi bounces down the path in nothing but coppery skin, body piercings and might-be-there-if-you-look-really-hard bikini bottom. Fawnlike in looks and nature, she is mesmerising and mesmerised, inquisitive and skittery, wide-eyed and life-inhaling. In the wild, she would have been devoured long ago, but here at the farm where's she's volunteering to gain her 2nd Year Australian Work Visa, she's been banqueting for two months.

My sun protected, long sleeved arms turn signalling pistons. I chop and point like I'm bringing a Hornet in on an aircraft carrier. I make slow-mo, silent words with my lips "Go b-a a-c-k ! Go b-a-c-k!". I will her with rotating lighthouse eyes to turn left to cover, to cover up. But the Disneyesque deer from Denmark is on her own spectrum, on her own path and on her way to the pool.

Unfortunately, I'm headed in that direction too, followed a few steps behind by twenty-six members of Australia's esteemed Country Women's Association. The ladies are on a tour, they're interested in learning more about native bees and the solar ovens we use to melt beeswax and bake slices in.

As far as I'm aware, Danish anatomy is not meant to feature on this tour.

The legendary women of the CWA have achieved massive amounts for rural families. They've lobbied, fundraised and organised since 1922; they've welcomed strangers, protested CSG, and breathed kindness and life into country towns the nation over. That's why I'm nervous today. Because they're here, and in my entire life, I've never made a scone, am a first-generation peasant farmer learning as I go, and I'm worried they'll judge me, my efforts and Bambi's might-be-there-if-you-look-really-hard bikini bottom as rigorously as they judge the fruitcake section at the Royal Easter Show. Oh dear, they might even judge me: fruitcake!

Judgement, etiquette, long skirts...I'd like to avoid you meeting pierced, soft, porn.

I stop the procession sharply, point my orange and white sun brolly to the sky and declare, "this way ladies, to the beetroot!"

The rallying cry reverberates around the valley setting off bleating goats, whinnies and the tinny whistle of a hearing aid. The invocation of Beta vulgaris – aka beetroot, CWA Plant of the Year – enables me to wheel safely back behind the crumbling stables, reverently spinning ladies like turntables as I go. A clash of civilisations and never-to-be-lived-down rural gossip narrowly avoided...again.

There have been lots of near misses lately and my face is the colour of beetroot from the extended flush of exertion, exasperation and exhaustion. After a decade we should have this farming life sorted, our debts paid, the chance for a weekend off and time to smell the roses. But the goats pruned the blooms, the bank said no to extending the overdraft and it seems everyone's on holidays but us...which of course is a fait accompli when your business is a holiday spot.

We didn't plan it this way, but through much fault of our own, we did make it this way, that's why only the good vibes and images find their way to the Internet, while the tough stuff stays for internal circulation only. But it's not just us.

Farm porn swamps Facebook and Instagram as though the ghost of Hugh Heffner is riding the tractor. Luscious herbs are tinted and cropped to beckon. Soft-filtered, oinking-porcines have their abattoir future edited. Buxom heads of broccoli bulge out like centrefolds.

"Look at this!" I say to Andrew, staring at images of perfect produce, funky model farmers doing yoga atop utes and immaculate tiny houses claiming to have been built on equally tiny budgets with no big trust fund, corporate salary, sacrifice or sweat. "That can't be true."

Farm porn shows no cabbage moth destruction, just artistic silver trails across yet-to-be destroyed leaves. The pictures of braided garlic – shimmering like Rapunzel's hair – betray no hint of the Rumpel-stiltskin-hunch the farmer (aka: me and countless like me but without the western privilege) earned during nine months of weeding, four hours on YouTube learning how to plait it like a pro, and midnights of braiding backache.

Farm porn doesn't let you in on the fact that there's a difference between dirt poor farm workers, dirt poor striving farm owners and dirt rich entities who have the money to buy water licenses in the current market and who's shiny corporate or family dollars can bankroll shiny feeds of camera kudos through Kumbaya, Kundalini yoga and kickboxing in front of kumquats.

Smoke. Mirrors. Photoshop.

Show-offs. Me-too's. Wannabees.

Me.

Some of those sultry, soft-focus images of alliums, brassicas and ruminants have been uploaded by dirt encrusted, beeswax burnt, turmeric stained fingers that look surprisingly like my own. Guilty, your Honour. I can't risk turning off mainstream customers by showing the grim reality of farming life...that if you have livestock, you'll have deadstock, that if you grow things, things might grow on them – and out of them.

The simple life, hijacked and hyped. Framed and filtered. A bit like my real life.

My favourite filter is the happy face. This is the one I have trained myself to put on. The last fourteen months I haven't needed it at all, thanks to the best run of Wwoofers in the history of the farm. One after another they just gelled. Ben and Charlotte, the chef and the cute heart; Roo, Tash and Rosa the English wanderlusters; Nicolette the gorgeous Dutch teacher who taught us about kindness and grace, and showed us we weren't the only people who cared or could get the job done. Then there were the British crew, captained by Jolly Jake, a tattooed mountain of a lad who needs to be sent to all the world's trouble spots to get them giggling again with his infectious, never to be immunised against laugh.

Grateful for their time with us and reminded of all the other awesome people who have come to help, we have no words, just swollen smiles and hearts, kids with grins from ear to ear and animals who have become accustomed to sleeping in hammocks, being entertained on balconies and sung to at sunrise and sunset.

That's why the comedown with these latest backpackers has felt so harsh. For more than a year we've been living with competent, compassionate comrades which meant we took even more projects on, produced more, did more, gave more service to customers and guests. Even with the seven-day-a-week work, it had felt like we could achieve anything. Blinded by the fun and freewheeling of it, we never realised that just because we could achieve more, it didn't mean we should, or could continue to. We'd been living the dream but now we were back living with the enemy: enemies of common-sense, enemies of joy, enemies of mental health.

I've always scoffed at the contestant's reactions on Survivor and Big Brother type shows, believing that when you know you're only with the other players for a short amount of time, you can control how much the situation and people bother you. Plus, we've done this before, more than a hundred times. Or is it a hundred and twenty? One hundred and thirty? Ups and downs. Swings and roundabouts. Whatever, I'm now in Survivor mode, stuck in my own Big Brother house and needing a happy face filter to survive the perfect backpacker storm.

Valentina is Brazilian, so volatile she generates her own weather patterns. She broods, she stews, she hands out dirty looks like parking rangers do fines. Her mood since arrival has felt like the oppressive humidity of a tropical low and with each passing day the barometric pressure has kept dropping. Where ever she walks, ostrich-sized egg shells rain down and I've been spending weeks avoiding walking on them.

Valentina is travelling with her manservant Gianni, an Italian without passion, an Italian without sauce. They began their three months at the farm at the same time as Trina, an American, groomed from toddlerhood to be a pro golfer until her shoulder putted; and Bambi, the Dane, so ethereal, so other worldly, I look to the planets each night to see which one she might have dropped from.

If each of them had come on their own, or been here with a different set of Wwoofers, it might have been tolerable, but the combination of this specific four has Andrew grinding his teeth in his sleep, and me spending the nights listening. But today there's so much to get done and so many bills to pay, I'm determined to keep my chin up and my blood pressure down. There's nothing like baking day to get feel-good aromas swirling.

The slice we're making is from my Mum's secret 1956 recipe. It makes entire bus groups, not just the diabetics, swoon with its sweetness and spice. Tourists and even the CWA ladies beg for the recipe and repeat customers snap it up at markets as soon as it goes on display. We make batches of it each fortnight in the busy season and cook it outside in the solar oven; unless the weather's bad and we need to use the inside oven, the door so wonky it needs a stool jammed against it to keep it closed.

Valentina and Bambi are in the kitchen awaiting instructions, Valentina thunder-clouding the chair at the kitchen bench while Bambi floats out of lightning range near the fridge. Bambi got on the bad side of Valentina within days of arriving by smiling, being nice, and wearing not much in front of Gianni. I'm not sure what I did.

"It'd be great if you guys can cut up the dates and walnuts so we can make more slice," I say to one stormy glare – Valentina's, and one vacant stare – Bambi's.

I give them cutting boards, a huge stainless-steel bowl filled with twelve cups of fleshy dates, and a slightly smaller bowl heaving with crinkle-edged walnuts.

"In half or quarters would be great," I say, showing them the approximate size I'd like.

I show them where the twenty-litre bucket of flour is, and where the honey, brown sugar, ginger, eggs and butter are, and talk the girls through the recipe. Satisfied they've got it, I leave them with the ingredients and their mutual distrust of each other, to go pack the ute for the market.

On the way to the shed I run into Andrew, Gianni and Trina. Gianni looks like a piece of spaghetti that's just been exposed to water: tall and thin – but saggy. His shorts sag below his waist, his undies sag out the top and the way he half-cooked-pasta's around the farm makes me want to tong him into a wheelbarrow and give him a lift. He's a man needing reinforcement, as does anyone in the vicinity of Valentina. Trina is the opposite, perky and preening, forever gazing at herself in reflective surfaces and expecting praise and a 'good shot!' each time she puts one foot in front of the other. Andrew's teaching Gianni how to use the ride-on mower. Trina's pretending to listen but is focussing on adjusting the wing mirror. I give them a wave as I head to the shed where racks rise to the roof, housing bottles of habanero honey, honey mustards, candles, rosella cordial, teas and more.

While packing the ute – a veritable jigsaw puzzle of a task requiring a degree in engineering and weightlifting – I think about the fine line with the whole volunteer for the visa work program and how I'm not quite a boss, not quite a mother, not quite a friend, not quite a host...and often, not quite good at this. And I'm reminded of it an hour later when I walk back into the kitchen.

Valentina and Bambi are cutting dates at twenty paces, glaring at each other like it's a western showdown, and literally cutting, with scissors, one date – and one walnut – in half – at a time.

Seriously? I am upset that they can't seem to get along, but I am more bewildered that on a day when there is so much to do, that they, women in their late twenties, women of the world and from the other side of the world, are using scissors to cut dates.

I wonder how they have no concept of getting twenty or forty sacrifices on a cutting board and cutting through them quickly with a knife. I wonder what brought them to decide on the one snip at a time method...are they imagining they are hairdressers to the stars? Snip. Do they feel so relaxed they think we have all the time in the world to coif a date? Snip. Blow wave a walnut?

I want to yell: "we're not trying to make a bonsai slice here girls! We're not trying to create an artistic piece that will feature in the Guggenheim or at Heston's or be snapped on someone's scalp at the Oscars!"

Do they have no understanding of how hard and fast we have to move to keep this little farm viable? Or that at the shops at 6am this morning I went red with embarrassment when my credit card was declined on the trolley load of double length toilet rolls I was buying for all their if's and butts?

Do they not realise my reaction to their whinging and whining is turning me into a grumpy, old-before-my-time person oozing cantankerism when I want to ooze enthusiasm and be c...a...l...m like an earth mother?

Do they not realise I know it's all my fault for complicating the simple life, that I'm bad at budgets and that though I want to be grateful and gracious for their help, if the scales tip from fair exchange to fail exchange in my incredibly short-sighted eyes I become sour and dour too?

And how do they not realise I'm particularly triggered today because another farmer said some of our miniature cows looked thin and I couldn't think quickly enough to tell him, 'that's because we're not fattening them up to kill them like you – that we just want our cows to be regular, fit, Michelle Bridges full-body-transformation-type-cows who will live long, non-obese, activated Pete Evans lives glowing as they digest grass that they turn into manure that turns into vegies and nuts.'

Nuts!

"Thanks guys," I say. "Maybe do it like this for the rest." And like a serial killer, I show them how to bowl cut multiple dates and walnuts at once.

We're finally at the stage of smoothing over and smoothing the mix into baking trays when Gianni comes in for his two-hourly feeding. The lad is keeping his spaghetti shape in tip top condition, though for what I'm not sure as he doesn't seem to expend much energy on any of the jobs we give him. I get out of the way so he can make his banquet and head out to see Andrew.

He's grinding his teeth, he's awake, and the story spills out. "So, I taught Gianni how to use the mower and he's been going for two hours and runs out of petrol. I go out with him to refill it and I say, "Gianni, I asked you to mow this area." And he says, "I did", and I'm looking around and the grass is still this high," he holds his hand to his thigh.

"I heard it going though," I say.

"I know. Two hours! Two hours of petrol! But he hasn't bloody put the blades on has he!?"

"He didn't notice?"

Who doesn't notice the smell of cut grass, the look of cut grass, cut grass sticking to your boots? Who doesn't notice long grass versus short grass? Fashionably cut grass? Oh yeh, I remember, another three backpackers who have stayed with us over the years.

"He's just run over it."

Andrew shakes his head, his teeth making that grrrrr sound. "Then I find out Trina melted the bench in the cottage with her hair curler and then I saw her feeding out forty dollars of hay to get the sheep to move through the gate without her having to go in with them. I asked her about it and she's been doing it every day for a week."

"We're going to have to say something."

"Yep."

When we open the kitchen door, the Nutri-Bullet is whirring at full speed making smoothies. Next to it I see the drained bottle of native bee honey. The one jar we get per hive, per year – and only after I spend three hours harvesting it, picking out each little bee from the liquid so they don't drown. The one jar of stingless bee honey, so renowned for its taste and medicinal qualities, so rare and precious, I had pre-sold it with five from our other hives, to celebrity chef and bush tucker wonder Jock Zonfrillo.

"Wha-what are you doing with the native bee honey?"

No one answers.

"The honey...you're using the native bee honey."

"Are we?" Asks Bambi dreamily.

"You didn't notice? Different type of jar? Tape around the top to keep it sealed? In the back fridge at the very back? Runnier? Different taste? Not one of these honeys?" I say, opening the pantry where regular honey, in round jars, hexagonal jars, teddy bear jars, big pots and small pots, huge tubs and hippopotamus-sized tubs takes up one whole shelf and half of the next one.

"Oh...yeh...we did think it was a bit different," says Trina.

"It's so good in smoothies we've been having it all week," says Bambi.

I am Mt Vesuvius with a blockage. I am a rib cage full of lava with nowhere to spew it. I am the angry face emoticon trapped in internal hell...because I've been here before with other backpackers and it's my fault I didn't label and hide it better, it's my fault I didn't show them how to use a knife, and it's Andrew's fault he thought Gianni would listen to instructions and notice the smell and look of cut grass versus hit and run grass. It's our fault and choice we run this business and crazy life the way we do.

Andrew, the calm one, begins. "It seems like this isn't really working. That you guys aren't all getting along, that you're not really into being here, that the animals aren't your thing...that maybe we're not the best hosts or experience for you..."

Then it's my turn, "...we totally understand. We know it's hard work, we know we're hard work, but no one is forcing you to stay. This thing is volunteer, you don't have to be here. If you're not enjoying yourselves, we want you to feel free to leave and find a different farm. We really don't mind, we'd prefer you were happy."

Silence. Everyone takes a moment.

"I definitely want to stay, I love it here," says Bambi.

"I have nowhere to go," says Trina on reflection.

"We are staying," thunders Valentina on behalf of her and jelly-like Gianni.

Silence. Andrew and I take a moment.

Stress, it's down there in my belly, boiling and roiling, forcing its way into the fissure that is my throat, then out my mouth. It's something that's been brewing, a culmination of living with more than one hundred backpackers, the good, the great and the unbelievable...I finally say what I've wanted to say just a couple of times in the past:

"Guys, this is our entire lives, it's just 90 days of yours... so if you're staying, it would be great if you could make an effort to try and fit in with us – rather than the other way around."

My happy face filter slides right off, and in an Insta-minute, by speaking my truth, I'm beaming again with my real one.

* * *

To get your hands on the recipe for the farm's secret recipe slice see page 253. Tip: to save time when cutting the dates, use a knife rather than scissors.

FARMERS' MARKET FATIGUE

*I*t's 3.45 am and I steer out of the farm, four Bridgestone tyres weightlifting hundreds of kilos of glass with honey barrelled inside. In the dark of the cockpit I crunch carrot sticks and snow peas, the mechanics of the mouth keeping the brain awake. At the Bulahdelah Bypass, half an hour south, it's time to crack a can of quick and easy cola – enough caffeine and bubbles to keep me awake but never a sip sooner than here – the Newcastle City Farmers Market is still more than an hour away and I don't want to toilet at the scary movie highway rest stops along the way.

For that first half hour, I berate myself for drinking from the aluminium of multi-nationals, for that first half hour I hate I'm habituated, but for that first half hour I know I'd hate it more if I were roadkill.

Truckies call it 'white line fever', the drifting disease of the mind as you follow hyphenated road markings for hours - - - not needing to touch the brake - - - drifting lanes, half asleep - - - drifting. I crunch, and sip again – to avoid that.

I think of my old city commute – less haul, more harry. Stop. Go. Stop. Go. The white lines invisible beneath the exhausts of others flocking peak hour. My Mazda 121, my little bubble, easily carried the tools of trade: one regular-sized cerebrum, no heavy-lifting required. I would risk peering at other shufflers: taut faces, tight knuckles, determined indifference, slog. I'd glance in the rear vision mirror and see myself: road worn from city life, spirit bald and with one thigh muscle so strong from the trigger foot above the brake.

Drift - - - and now here I am again, commuting. I keep my brain awake by calculating how many hours I've spent on farmers' markets. It's too difficult to work out when I try to add in the hours making the balms, the soaps, the growing and drying of herbs, the harvesting of honey, the picking of fruits, the labelling of jars...so I work out an easier formula.

Time spent packing products into boxes + ute loading + average drive time + stall set up + market selling + pack out + ute unload x the number of markets at places as far and wide as Warriewood, The Entrance, Gosford, Gloucester, Port Macquarie, Nabiac and my destination today, Newcastle.

An hour of monotony and maths has passed on the journey and the highway rumble strips kick in – the raised painted lines ker-thunk ker-thunk me awake so I don't take the roundabout at full speed.

Three thousand eight hundred and seventy-four hours.

Three thousand eight hundred and seventy-four hours of my life doing farmers markets.

That startles me more than the rumble strips. Not even half way to Malcolm Gladwell's 10 000 hours of expertise, and no longer feeling the burning desire to get there. What do you do when what you once loved becomes a chore?

During the week, Newcastle roads choke like the exhaust of the coal it exports to the world, but at 5.15 am on a Sunday, the only thing in front of me are deserted streets and the Navara's headlights. Twenty minutes later I await the green arrow to swing into the Newcastle Showground, then inch along, avoiding tailgates, bent figures and the moving glow of lit cigarettes. It's an indoor-outdoor market and I need to get to the back of the big sheds, past the skip bins and reverse to the second roller door. No one is here yet so I can park close, unload, then get out of the way. Then, more headlights.

"Bloody in the way again," gruffs the lettuce man, jesting and jousting, comedic and cantankerous, revving his engine and ribbing like his greens.

"Yeh, yeh," I smile outwards. I'm in my unload place, respectfully less than a finger's room between my ute and the side of the shed. He can access his spot reserved through more years of toil than me – but he still wants me to know it took him an extra turn of the wheel to get in. I've been making the fortnightly trek here for years but I'm still very much the junior.

"How do I get a prime park around here?" I ask him playfully, knowing the answer is 'a bad back and a heart condition'. I wonder if it's worth doing my time, my spine, and another three thousand eight hundred and seventy-four hours – for a closer park?

I transfer the weight off the Bridgestone's to my shoulders, my arms and my fledgling hiatus hernia. I unload, backwards and forwards, till the tray is empty, till my muscles are too. I move the ute then reach down to the passenger side for the cash tin, the fridge bag with the produce testers, and my water-filled Klean Kanteen and walk in side by side with Jeffrey of the Olives who greets me as usual with his wry smile and a day's supply of warmly delivered X-rated jokes.

I unfold the tables and begin draping table cloths down to the floor to block the view of the boxes behind.

"A millimetre out on this end," teases Dukkha Dude. "You should bring a level. And your trendy signs…been to shop window dressing school have you?"

"I wish," I say, as I put out my signs with their chic-cardboard aesthetic, a style achieved not by design but through determined recycling and budgetary constraints. They're looking a bit worse for wear but I don't have the oomph to re-do them, I feel I'm fading, a bit like them.

Beneath the corrugated iron roof, this 3m x 3m concrete-floored space will be my pen till 1 pm, so I make the 6 am hope-it-all-comes-out dash to the toilet: past the sides of goat, the microgreens of Lilliput, the lemon wedges of biodynamic cheese; through steamy clouds rising from Dutch waffles, cinnamon chai's and grass-fed sausage sizzles; ducking and weaving as marquees rise, trucks reverse and stallholders jostle, while beside them utes root into the concrete roadway magically blossoming bananas, blueberries and blood plums.

Then back via the alley of bromeliads, begonias and buxus; down the aisle of Isa browns, Australorps and silkies; come around cauldron corner where kernels pop-pop-pop ready to be doused in butter, cinnamon, sugar and salt, the aroma following me all the way back to my pen.

No trading is allowed till 7 am, but the faithful hungry are already here. They push before them patterned fabric trolleys, eskies on wheels, laundry basket rollers and arms lassoed with recyclable bags. They know exactly what they want, and if it's not you, they canter past, blinkered like horses. In the early days, I'd be so excited if someone even glanced my way, but now as the mob canters past I feel I'm beneath their feet.

The early risers head first for the French pastries, the fresh cut proteas and waratahs, the loaves of organic rye, cartons of free range domes and fragrant organics of basil, coriander and parsley. They trot these to the car, returning for kilos of bananas, buckets of tomatoes, peaches, mandarins, and overflowing saddlebags of silverbeet and leeks. Next, they ride for the refrigerated salmon, featherless duck, pastured pork, steaked steer, salty haloumi and milk creamy as yoghurt. Corralling the fresh produce in the back of their car they slow to a walk, wandering the trails of preserves, powdered spices and infused olive oils, the paleo bars, the fizzing kombuchas, the pomegranate balsamics. They work in teams, pulling their drays and grazing at the tasting plates in front of the stalls. If you whisper 'hello', some shy away, some bolt, some stand their ground and prick their ears. Some even come to you uncalled.

They say things like, "where's your honey from?"

"Ironbark's from our bees at Nabiac, the Yellowbox and Applebox from our friend Daryl, he's a migratory beekeeper so takes the bees to the trees. Would you like a taste?"

They snort things like, "all honey's the same."

All horses are the same, until you ride a Shetland, a Thoroughbred, a Clydesdale. All wine's the same until you sip Cabernet, Chardonnay, Chambourcin. All cheese is cheese until you mouth Blue, Cheddar, Camembert.

They tell me things like, "I'm getting bees," and stand rump to other customers for forty minutes, stampeding me with questions.

They school me in their knowledge of my livelihood, "a lot of the supermarket honey is blended, you know? Did you hear about that company who got busted for selling corn syrup as honey? Lucky we don't have that Varroa mite here yet, and bloody Monsanto with their chemicals! There's a lot of imported Chinese muck being blended and called Australian – can you believe they can get away with it!? The bees are all dying, dying! America only has 'em because they fly millions of replacements over in QANTAS cargo holds every week –- they need to pack them in dry ice so they don't overheat the planes. It's like going to war in those Californian almond fields, they never come back. Poor bees."

They ask, "you don't feed the bees sugar, do you?" Never. "Are you organic?" Yes, but we don't own the 5 km of land around our hives so can't be certified. "Do you heat your honey?" No, it's raw, like the bees have it.

They complain. "It costs that much?! Honey's so expensive these days. I get mine free from my neighbour. It'd be cheaper if you put it in plastic not glass. What is it – gold? Can you do it cheaper than the other honey sellers here, I don't like theirs as much as yours. Would you take one dollar off? Can you give me three for the price of two?" I can't back away but feel forced to back down, then I silently paw the ground, remembering Rod the Beeman's mentoring of me, "tell them they can have it free...if they come do the bloody work!"

They say, "I'll be back later to get some." Which of course means they won't.

And with noses in the air state things like, "I don't like honey." Until I tempt them to try.

The dip of the little stick into honey pot one, *Mmmm* – the taste of sunshine; the dip of another into honey pot two – *Ohhhh*, so sweet the mouth waters; and into honey pot three, *Ooohh* – caramel, maple, wood-smoke and treacle. "I never realised honey could be that different, that good! I'll take a jar of all three thanks."

They say things like, "thank God you're here, we love your stuff, can tell it's made with love!" They say, "we've been to your farm, we loved it." And a few say, "thank you so much for making the effort, for caring for the land and the bees."

These few see the richness of the earth, presented on my table. These few see the mining of muscle and heart; nature and homo-sapien entwined. But it is too few to stop the slow fracking of the balance sheet and optimism; too few to release the reins reefed by others.

By midday the herd thins and stallholders slouch. The water-coolers of this office are the doorways of refrigerated trucks, the poles of marquees and the corner stalls of the sheds.

"Not as good as last week," says Stanley of the condiments.

"Tracking down, year on year," says Carol of the sauces.

"Bloody Aldi," says Drizabone-wearing of the grass fed.

Then the boardroom breaks out:

"More marketing would be good."

"Lower stall fees would be good."

"But you don't want to cross 'em," someone under their breath, the threat of being banned by any market manager an ever-present concern.

I look around at the harvested tables, windrows of empty cartons, farmers earned and spent. Husks, hulls, empty punnets. Office workers of the earth, minus the luxury of nine to five. The fresh air and wide open promise of farming compressed into sheds and marquees, fenced by seven day weeks, hours of dashboards and patrolled by bankers. The country feeds the city, the city gorges, the country starves.

My mind drifts - - - my heart gallops - - - after years free ranging, the brumby feels she's commuting again. When did the joy escape? Did I leave the gate open? Will it come back if I call?

* * *

IF YOU'RE EVER EXHAUSTED AND NEED TO WHIP UP A QUICK DINNER, there's a recipe for Honey Haloumi Fry on page 245.

FAT MAX AND THE FORTY THIEVES

at Max is a black and white Dorper sheep and channeller of Po from *KungFu Panda*. He's so fat everyone thinks he's about to drop twins.

"You sure he's not a Maxine? You sure?" asked one insistent guest.

We once found him at the end of the road paddock in a slight dip, four legs pointing skyward, unable to right himself because of his almighty saddlebags. When we got to him, he had a casual look on his face, like he'd just discovered the joy of sheep sunbaking. He protested momentarily as we rolled him out, then asked for a cocktail.

When professional sheep farmers visit, they comment *he's a strange one* – like most of our flock – because they waltz right up for chin scratches and cuddles. And they can't believe their eyes when Andrew whistles, and forty of them bolt across three paddocks to be at his side.

"I have to use a fifteen-thousand-dollar dog to get them to do that," says one Merino breeder, shaking his head at the sight.

We originally purchased the Dorper flock – a South African breed combining Dorset Horns and Blackhead Persians – to help organically control fireweed on the farm. Whereas the fireweed of the Northern Hemisphere is the beautiful, magenta-flowered *Chamerion angustifolium*, a great source of nectar for bees, the fireweed of the Southern Hemisphere, *Senecio madagascariensis,* is yellow like a daisy and toxic as hell. It arrived in Australia around 1918, in the ballast of ships trading the routes out of Southern Africa. The serrated green leaves are the only thing to hint of its danger and just a small infestation on a farm can erupt one million seeds per hectare in a season. The plant itself crowds out grasses, clovers and herbs, so animals have little choice but to eat its poison.

Horses suffer brain damage and blindness from the plant's pyrrolizidine alkaloids, while cattle lose weight, get spurting diarrhoea (known as scours), and eventually die. Pleasant! The key ways to control the overtaking of your paddocks by this weed so noxious it's been declared an evil of national significance is to (a) use the child labour company a.k.a. Lucy, Jack and Rosie, to pull it up, well-gloved of course because the toxins can also be absorbed by the skin; (b) spray even more toxic herbicides across the farm, or; (c) use sheep to eat it – because they're up to twenty times more tolerant to the alkaloids, and can literally nip it in the bud.

So that's how Fat Max came to be with us, the son of one of the first mothers we'd bought to tackle the fireweed. Instead he's tackled us, with a therapeutic devotion only another species can give.

I'm admiring him now.

As I collect honey from the bees, he's positioned in the shade of the ute, leaning against the back tyre, watching me at work. His jaws are perpetually in motion: masticating, mumbling, trying to form words? He's such a cool cat, and he knows it. But he's a nice guy too, never butting or pushing, a steady, kind, sheep-couch who you can sit with for hours and ponder the universe.

I heave the last of the honey supers into the tray of the ute. Six boxes need to be extracted today and I need to get started.

"Come on Max, you're going to have to move mate."

But he can't be bothered, so we sit there together, ignoring the bees who are ignoring us, ignoring the work that needs to be done, and enjoying the not-doing-anything moment of just being.

Those moments are rare at the farm, cherished. When I finally drive back to the farmhouse I learn Andrew's Mum is in the Intensive Care Unit in Sydney and he's heading south to be with her. It's a family emergency and he needs to go and I want him to go, but I dread him being gone. It's just that things always go pear-shaped when he's not here: wild Wwoofers on Tinder dates with drunk locals, dying chickens, goats in trees, impossible guests, unannounced bus tours...I finish the day with a blur of honey spinning, relying on Sienna – a capable return Wwoofer – and Rosie the animal-whispering ten-year-old, to do the afternoon animal rounds.

We convene in the kitchen.

"500 000 bees," says Rosie.

"Check," says Sienna.

"46 chickens," says Rosie.

"Check," says Sienna.

"28 goats," says Rosie.

"Check," says Sienna.

"22 cows."

"Check."

"2 horses."

Check.

"6 alpacas."

"Check."

"33 sheep."

"31."

"You sure?"

"That's what it says on your Dad's list."

And while I cook veggie enchiladas, they start re-scouring the road paddock for the sheep. It's dark and the quadbike lights beam just far enough in front of them to avoid ditches and springs, while they use another torch to sweep the grass.

Andrew instituted the afternoon animal count to ensure no one was ever stuck in a fence, ill, giving birth, foxed, waterlogged or needing help in anyway. His attention to detail had saved numerous lives over the years. But when I hear the quadbike returning and Rosie calling, I know it hasn't tonight. I head back out with them.

The torch light shows up tufts of hair where the dogs have latched and snatched, gripped and yanked. It's a grotesque trail leading to our beautiful Fat Max.

A huge grey watermelon billows out of his back and his eyes are soft and dark. I look into them, startling as I realise they're gone – his sweet jewel-like eyes stolen already by a feathered nocturnal thief.

Rosie trembles and I hug her away, she's not allowed to watch M-rated shows yet, but her life involves excruciating close ups into life and death, beauty and tragedy, love and loss.

I drop her home and get a tarp to cover him, there will be no more desecration of this amazing being. Sienna helps me drag logs to anchor the tarp, we'll bury him tomorrow, then we grid search by torchlight for the missing ewe. There she is, in the long grass, her head left intact but joined to a rib cage picked clean and white like the bare ribbed frame of a boat.

Rosie sleeps with me that night, and when Andrew returns we bury Fat Max in a big fat grave and scatter big fat white rose petals and shed big fat tears for the sheep who made us think, for the sheep who made us feel, and for the sheep who made us happy.

It's the second dog attack we've had in the last few years and when the sun goes down it's hard to sleep – which is a problem cause that's when you're meant to.

It's a week later and the emergency lights in my brain are finally ready to dim, I'm about to dissolve ethereally into my pillow with no thought of snarling cross-bred dingoes or neighbour's dogs having a go, or Stephen King Cujo's, when I hear it: *Clank*.

"You got a goat up in the dairy?" I call up the corridor to Andrew.

"No."

"I think something's up there."

"It's called 'the wind'."

It wasn't the wind, it was a *clank*, and it's enough to flick my brain back on high beam.

"Sorry," I say, seeing him slumped on the couch after his own massive day moving honey barrels, trimming goat hooves and staking seedlings. "Can you go check?"

There's always something to go check here. Is that alpaca sleeping or dying? Why is that goat still bleating? Do the horses have their rugs on? Did we turn the hot water on for the guests? Is the shed locked? The shop? Is that grass green from fertility or has a pipe split? Can you hear that cow, does it sound right to you? Do we have enough slice for the bus group tomorrow? Did I pick all the stink bugs off the orange tree? Did Princess go into labour yet? Can you check where the stitch kit is in case her uterus comes out? Are you sure the honey tap isn't leaking? Are the chicken caravans safe in this wind? Are all the sheep and goats in the safe paddock...is it really safe? Maybe we should just go check.

And night-time checking is Andrew's domain. That's because he's never been one for horror movies; he's not expecting Cujo or Freddy Kruger, Michael Myers or Leatherface, not even Pennywise or an inappropriately positioned horse-head. He doesn't worry the Children of the Corn might be out on the lawn scythed up and ready to harvest him. No, a night-time walk around the farm, baton-like torch in hand, scoping the scary noise trail, doesn't rattle him at all. Me, I'll just wait right here, lock the door behind him and crouch in the dark with a serrated knife and phone – on silent of course – ready to dial for police, fire, ambulance and Ninjas if they'd just give me their number.

Ten minutes later he's back. I shine the torch in his face to make sure it's really him. Then shine it again to make sure no one and nothing's behind him. Only then do I unlock the door, let him in.

"Nothing, just the wind."

Whatever, I stash the knife under my bed anyway, and try and stop my brain from blazing on the pillow. My batteries finally go dead around 3 am.

Morning comes and I feel around the veritable arsenal under the bed. In addition to last night's knife I retrieve a steak knife, fishing knife, boning knife, carving knife, plus scissors, box cutter and home-made pepper spray...well what else am I going to do with all the extra chillies? I take everything but the fishing knife and pepper spray back to the kitchen, and make like a normal housewife with the cutlery draw.

Andrew, up since dawn, has already made himself his signature omelette. This last week since he returned from hospital and with his mum in recovery, he's been determined to get a good start on the day. He starts before six, attacking the mammoth to-do-list he keeps in the tiny ring-bound notebook in his pocket.

"You seen the whipper snipper Jack?"

Jack's not much into the guest or animal side of farming, but give him anything with a blade – machete, saw, scythe, mower, whipper snipper – and he's the most enthusiastic worker on the farm.

"Near the tractor."

"I looked there."

Clank.

The cereal in my mouth suddenly turns to cardboard and I start thinking of a different type of fox, the two-legged kind: "Might want to check if anything else is missing."

Twenty minutes later, hands-on-hips, Andrew returns. "All the power tools."

Stealing has happened in the area before. At the time of white settlement, the land was inhabited by the indigenous people of the Worimi and Biripai tribes.

I met one of the Biripai elders once at a Landcare gathering. He carried with him a ball of twine and gently teased out a strand until it was 15 cm long. He was pointed, when he pointed out, "this is how long white people have been here." Then, using his fingers as the spool, he continued to unravel the ball until it lay from one end of the carpark to the other. Gently he spoke now, almost a whisper, "this is how long my people have been here."

At the time I'd felt such shame, shame at the vast loss these people have suffered, not just in land, but cultural richness and identity. I felt shame at having bought and fenced land that hundreds of years ago they'd been able to roam freely. And I felt guilt, wondering if what we were doing here, trying to regenerate the land and encourage visitors to make a connection with it, honoured in anyway nearly enough the original custodians. And I almost laugh now at the outrage I feel that we've been robbed, when their losses have been so incomprehensible.

What goes around, comes around. And someone had been going around our farm last night. What creeps me out the most is it was a human behind the *clank*, a human walking in pure moonlight but on legs of darkness, a human Andrew could have come face to face with. What then?

We report the theft to the police, not expecting much, but a day later we get a phone call.

"Did you also happen to lose some saddles and bridles? An incubator? A compressor?"

Clank. That's what I'd heard, the sound of a stirrup banging against the corrugated iron walls.

We hadn't even checked the tack room, but gone are the saddles we need for farmstay pony rides, gone are bridles, alpaca leads, and the bright blue and orange incubator that's brought hundreds of gorgeous chickens into the world. The blood in me, tepid and calm till now, begins to bubble. How can't the robber see we've worked so hard to get these tools of trade? That we're not in a position to just pop out and replace them? Why shouldn't he have to work for them too? I'd be happy to show him how and help skill him up.

"There were six of them, but don't worry, we've got 'em talking."

Six? Six?! Six people so close to our bedrooms at night? Six people Andrew could have walked right into?

"They saw your husband's torch and took off or they would have kept helping themselves." I feel sick. "It was a couple of local fella's, couple of their mates from out of town."

Now I want to heave. Do we know these people? Their parents? Have they been here as day visitors and scoped the place? Are we under watch? Do we need to man-trap the place with snares, camouflage pits and pressure-release weapons of medieval torture?

Or is it my fault? Only last week I'd put a call out to the Universe for help simplifying our lives. I hadn't realised at the time her idea of decluttering was to send us six petty crims and a pack of wild dogs.

"That's not what I meant!" I yell into the air.

Over the next few days our stolen items begin to trickle back. The detectives reverse into our carport every second day, popping their boot and asking, "is this yours? This? What about this, is this yours too?" One day they pop the boot and under one of our retrieved saddles lies a thick bundle of fresh cut marijuana.

"The saddle, not the green stuff," I say, and they laugh. I do too, but choke back my tears.

There's no Fat Max in there, nothing with warm eyes, sturdy chest, kind and trusting heart. He stays stolen, never to be returned. Fat Max isn't coming back, and of all the things that disappeared this week, it's him that we want.

* * *

I STOLE THIS RECIPE FOR SPINACH AND RICOTTA DUMPLINGS FROM MY sister, it's a great comfort food and it's on page 249.

REBOOT GUMBOOT

*T*he Port Macquarie Real Food Market is held on pavers just outside the glass-fronted Port Central shopping centre. It's not your typical venue for a farmers' market, but it's brought a buzz to the area on a Tuesday, has lightened up the loiterers and has been attracting locals and tourists since its inception a few years ago. It's an afternoon event, so there's no need for the pre-dawn alarm, and it means I have plenty of time in the morning to hang with the animals and then dig fresh turmeric to bring along.

Physically, it's the easiest market we do, as we're able to park the ute right at the stall; and that's crucially important now, as I've hiatus hernia'd myself doing so much heavy lifting. I'm also as fragile as a crystal glass on a cricket pitch after a soul-bruising time with one of the last lot of Wwoofers to share the farm, a group who managed to be even more taxing than the last, and on a much deeper level...so deep, it felt like I was submarining through my own Marianas Trench.

Geez it's dark down there.

I've lost so much weight I'm wearing a belt to keep my jeans up. There's no lifting thirty kilo tubs of honey on my own any more, no lifting beehives laden with golden honeycomb, no running with wheelbarrows. The weight's come off me so quickly, and I've felt so sick, we've had to ditch the Newcastle Market – and most of our farm plans too.

The busker on the steps of The Glasshouse is singing, "I've seen fire and I've seen rain, I've seen sunny skies I thought would never end..." and I wince at the irony, because the last few seasons have left me thinking I'll never see the light again – and right now I'm standing under a tree being peed on by cicadas.

Why am I not surprised?

Because they did it last week too.

That refreshing sun shower on my face is a nature call, delivered by a gang of Greengrocer cicadas and a couple of their *Cystosoma schmeltzi* buddies, aptly known as the Lesser Bladders.

Incontinent cicadas are raining on my parade, sucking in so much tree sap that they and I both feel we can't hold on any longer. The planet, I realise, is taking the piss out of me and recycling it overhead; cicada urine just the latest in a series of things these last few years that have been coming my way – but not going my way.

Most momentous. Most unexpectedly. My Roddy died. A simple four-letter clot.

Rod the Bee Man, my mentor, my insect safety net, my compatriot in all things bug for a decade. Rod, who would drunkenly fire a shotgun in time with the New Year's Eve fireworks; who would let a bee sit on his palm for as long as she liked; and who would drop whatever he was doing to do. To be. To bee.

Rod, who was warmth and kindness and optimism.

Protector and director.

Family without the blood.

Blood without the boil.

A bee without the sting.

An unbearable loss to those who knew him.

And even to those who knew him not.

Rod the Beeman grew wings.

Swarmed. Left the colony who loved him, the girl who needed him, empty.

...the potency of vacancy...

Then, there was Fat Max and the robbery.

Then, the government announced it was putting an end to the volunteering for the second-year visa scheme. The ten applications a week we got from backpackers evaporate to none. All the great reviews we've received online from past Wwoofers matter for nothing.

From across the country we started hearing reports of small farms closing, reducing their production, dropping markets. We've been undercapitalised from the start on this venture but have got through thanks to loyal farmstay and market supporters, amazing backpackers and happy delusion. But now we're gasping for breath. Serious, lung burning, feel like we can't breathe, just can't get the air we need g...a... s...p...s.

It's idyllic torture to be on a farm, surrounded by a breathing carpet of lime, a crisp curtain of bright blue and the stereo and smell-a-vision of life – and to feel you can't breathe.

G...a...s...p.

Our lungs collapse, we collapse, and so does the vision.

I can't lift, I can't rise to the challenge, I can't see where my internal or external cavalry will come from.

Then we get the annual insurance bill, it's risen again. It's hard to pay five figure bills when the bulk of the products you sell are single digit.

We weigh up the effort and expense for the risk and return, and the result is the last straw. The bale topples down from the top of the haystack.

Topple.

Topple.

Excruciatingly,

slow motion,

topple.

Nearly there.

The last straw reaches the ground. And we close the farm to the public.

With the help of our beekeeper friend Daryl, we've been able to keep providing a range of produce and honey to market customers. Here's one of those customers now, but she doesn't look happy and I don't think it's because she's being peed on by cicadas. I doubt she even knows.

"I tried to visit your place the other day," she says, glaring over the honey display, "wanted to bring this one." She points to her four-year-old who's happily leaving teeth marks in one of the beeswax blocks I have on display.

Won't be able to give that one away.

"I'm sorry," I say.

I've had to say sorry a lot lately. "Sorry, I haven't had time to make more balms. Sorry, I can't get to that market anymore. Sorry we're not open." Sorry, sorry, sorry.

Today, I'm running the stall with Minnesota Melissa, she's the very last Wwoofer through on the cancelled visa scheme; and she's basically a human balm after all the recent scrapes. Soft brown curls, soft-brown eyes, soft heart. She's the most diligent young hippy I've ever met, earthy and awed, connecting not grasping, hard-working not shirking.

"Can I help you with anything else?" asks Melissa, taking the customer's attention. She's heard me apologise a lot lately and steps in like a shield. But the lady doesn't want to buy anything, just wants me to pay for her disappointment, and walks off, dragging the little boy with her.

In her place come a steady stream of buyers for the turmeric. Years ago, when we first started growing *Curcuma longa*, no one recognised its bulbous rhizome or wanted it, but this year it's suddenly hot in the media, and triathletes to the walking-framed, buy bags at a time. It makes me happy that people want it because I know how good it is and how it's been grown in soil rich in organic manure and nutrients and love. I also realise I really want to give it away as opposed to selling it. It's a gift from nature to all of us and it feels wrong to charge for life. Why have we humans turned everything into a transaction?

"Just picked this morning, you'll need to give it a good wash," I say. "Be sure to add black pepper when you use it, that'll increase your curcumin absorption, that'll help with the anti-inflammatory properties."

I need to eat about a tonne myself to help take away the pain and guilt of farewelling most of our honey bee hives. Not being able to lift them, not being able to care for them how I always have, stung worse than they ever did.

My broodings are punctuated by ol' Noel the mandarin seller next door, who every few minutes ditties the crowd with his favourite calls to action:

"Buy from me someone, anyone, I want a new Lamborghini this year."

"The sweetest girl I ever saw was eating me fruit in the raw."

"All I ask is you don't squeeze 'em, save that for your husband."

"The only thing wrong with these strawberries is they're too cheap!"

As the flying foxes take in their thousands to the darkening sky, we pack the ute for the hour-long trip back down the highway to the farm.

"What are you going to do when I leave?" asks Melissa. "How will you do everything with no other backpackers coming through?"

"I think we're just going to do less," I say. "No more running after people, no more trying to be everything to everyone. Just live a simpler, more sustainable life, like we originally dreamed."

I don't say, but do think guiltily, 'we might have to rent or sell the farm, we might not.'

The headlights show me the road, but they don't show me the way...yet.

I wonder who I'd be without my uniform of farm boots, tee and cap? What's my identity if not the hardworking honey lady? What's my identity if not the Mumma environmentalist? What's my identity if I no longer try and inspire others to think about life, health and earth? What's it say about me if the care fatigue wins out? Does Al Gore secretly run his air-conditioner on turbo? Does the Dalai Lama ever get grumpy? Did David Attenborough ever harbour thoughts of switching to mining to make a buck?

What happens when you can't afford to dream any longer? Or when you fall out of love with the dream? Or when the dream no longer fits? Is this what it's like for a passionate teacher when they love the kids but can't face the parents anymore? For an actress who worked so hard for fame but who no longer wants to be recognised? For a suburban mum or dad who wants to shed their responsibilities, worries and skin?

Or what if the end of one dream is just the start of another?

"I think I'll play my 'om' tape tonight," says Melissa, intuitively trying to still my mind.

Aum. Aum.

I hear the chant in my head. The chant of Hinduism, Jainism, Buddhism, yoga goers at the beach and in meditation retreats around the world…a cosmic sound to connect, bring peace, share goodness. A sound, I've sometimes thought, that is the soundtrack for Westerners who don't really want to get their hands dirty, who don't really want to give anything up to make the world a better place…except a few minutes sitting, stretching and sounding. Ouch. Someone who I don't want to turn back into.

I've never taken the time to *Aum*, partly because the farm's gone from a dream to an obsession, a foray to a melee, a choice to a chore. It's like the scoop of ice cream that turns into a tub, the reading habit that turns into fourteen bookcases, the love of animals that turns into three hundred mouths to feed.

Aum. Aum.

I've become the cat lady of the eighteen-hours-a-day, seven-day-a-week simple life – the simple life on steroids – unable to part with the doing, the duty, the optimistically stubborn belief it will all work out;

that we can do all this, afford all this, and still find time to relax and potter too.

But how can we explore new things while still doing everything? How do we build ourselves if we're forever trying to build the soil and balance the balance sheet? How do we make the simple, sustainable life…simple and sustainable?

Maybe it's not all our fault though, maybe the earth has had a hand in it too. Maybe she's used us to renew her, not the other way around.

Maybe she owned us.

Maybe she's been the keeper and we've been the worker bees.

Aum. Aum.

Aum…hmmm… I wonder if there are professional Aum-ers? There must be if Melissa has them on her playlist. Is there a competition for them? Like Australia's Next Top Aum-er? The Aum Factor? Australia's Got Aum?

What if you want to be one, how do you go about it? Is there a school? A guru? Can I be one? Maybe that's what's next for me.

Maybe, I *Aum* on it…it's the searching and discovery that keeps the soul happy, that sets the entire being aflame. Not the getting there and doing it, doing it, doing it. It's the journey to *it*, not the *it* itself, that matters. And these last ten years we've certainly been journeying, and for the most part, on a trip to grateful.

The earth has given. And she's taken.

And so have we.

The next day I'm convalescing in the camp chair. I take off my gumboots, my socks. I look between my bare feet at the silvery reflections dancing on the dam. The olive coloured dam water is afire with sparkles, and the six-storey high ironbarks, leaf pompoms at the end of their branches, chant and shimmy like cheerleaders in the breeze. The solar pump sucks and pulls, and as the southerly sweeps through the valley, the swaying grass sounds like a thousand taffeta dresses at a dance.

I point to the tree with the large cone-shaped split, bulging resin from its trunk.

"Do you think it's healing or dying?" I ask Andrew, who's reclining in the chair next to mine.

We're in our private spot, down outside the canvas tent we erected last spring when I found I had nowhere to escape the mental cacophony of so many people in the house, the garden, the answering machine and my Marianas Trench. The tent's become one of my favourite places to just be, with the added benefit that the canopy acts like a giant nappy for the cicadas overhead.

"No idea," he says, and I have no idea either.

Whatever's happening, it just is. The tree will deal with it, or it won't. If it does, it'll provide nectar again for the bees next year and shelter for the birds who cluster in its branches. If it doesn't, it'll provide a grand scratching post for the cattle, and decades of decomposing bark for the soil.

Decomposing.

Enriching.

Reconstruction.

Life.

Epiphany.

This part of my life is just part of a cycle, not necessarily the end.

When we'd first arrived in Nabiac, it was as if the town was on its last legs, but over the years the closed signs on shops have given way to eclectic, energetic new ventures; the little farmers' market that could, did; donations to the community swimming pool finally saw it built and filled. Vibrancy returned with new ideas, energy, bright sparks, bright paint and community spirit.

That's what I want too. I want the vibrancy to return. I want new ideas and energy. I want my spirit back. But the only way to get it, is to acknowledge the unspeakable.

"Babe," I say unspeakably, "I'm tired."

He adds his own voice to the unspeakable. "I am too."

He squeezes my hand, for the longest time, and I think about plants and trees and the cycles they go through. Maybe I'm just like the old tree, weary right now but ready to renew. I just need to take a cutting of myself and nurture it to start afresh. A cutting that keeps my core and all of me intact, but enables me to renew, shoot ahead, reach for the sky.

Grow in wonder.

Bloom.

Oxygenate in life.

Andrew's still holding my hand as we walk from the dam up the few hundred metres of gentle grassed slope to the bee sanctuary. The trees have really rooted in now, no longer tiny seedlings but spreading arms of foliage and flowers, trunks thickening, oxygen replenishing. They're nowhere near their peak, but they're stretching toward it, and hopefully will for decades to come.

I send a wish to the carobs, the corks, the pecans and bunyas that they'll make it well past their first century. I send a wish that one day in the future when I try to wrap my arms around them, that I'll remember the amazing feeling of holding such tiny promises in my hands. I send a wish to the bees that they'll always find shelter and food here and that humans – centuries from now – will discover goodness in the soil, and themselves.

As we slide into the seats on the viewing deck, we take in the show.

The sun is setting in the West, turning the planet's lid of clouds an other-worldly ripple of bronze, pink and gold. From this angle, the farm looks as though it tips into its centre; a kind dimple of an amphitheatre capturing sunshine, rains and delight. Our laughing children, leggy teenagers now, clank and climb on the metal ribs of the cattle yards; goats bleat for human affection; lorikeets gossip in the gums; and the low, electrical hum of multiple bee species blanket the land with life.

The air is spiked with rose, lemon balm, mint; a mix of beauty, calm and zest.

I realise, we can't make it work the way we have till now, and we've made only a fraction of the difference to the planet we'd hoped, but I know we've come a long way from our old ways.

Though we've made mistakes, we've made efforts; though we've stuffed up we've also made stuff; though we've not improved our connections, we've made the connection.

We've made hay while the sun shined and made our family's lives a rainbow of unforgettable people, animals and experiences.

Over the last ten years we've made a life.

And I'm happy.

Because that was the dream. And it's plenty enough for now.

TIPS, THOUGHTS & RECIPES

Over a decade on the farm I experimented with and made numerous balms, was gifted recipes from friends and learned so much from so many. One of my main reasons for writing this book was as a way to share them. That's what this section is all about.

Please don't expect recipes of the quality of a celebrity chef and feel free to play around with them to make them better, all I hope is that you find something here that will be of interest or that you will enjoy. If you do, or you have a suggestion to improve anything, you can visit the website for this book at www.honeyfarmdreaming.com to share.

Finally, some of the ingredients in the balms can be expensive, and these recipes are for big batches. Either reduce the ingredients or get together with friends or your book club to pool resources, or plan to give them as gifts throughout the year. Always patch test the balms to check for allergies.

All the best to you and yours,
Anna xox

A LITTLE ABOUT FOUR LEAF CLOVERS

Personally, I still haven't found one! But Rosie is a gun at spotting them. It's thought that there is one four-leaf clover for every 10 000 three-leaf clovers, so, like fairies, they can be very difficult to spot!

The four-leaf clover is a mutation of the three leaf. It can occur due to a recessive gene or environmental factors (they are often found growing in poor soil). It can seem incredibly difficult to find one, but Rosie says the trick is to stand and look down at the clover patch, and look for square shapes, rather than looking at all the individual clovers. Once you've found one in that patch, there is a good chance you will find another nearby.

The four leaves of the clover are meant to signify: hope, faith, love and luck. In the old days, people thought finding one meant you'd be able to see fairies!

To dry a four-leaf clover, the best thing to do is immediately on picking, lay it flat inside paper towel or wax paper, then put it between the pages of a very heavy book...and then stack some more books on top! If you don't do it soon after picking, the clover will wilt and the edges will curl up. Leave your clover like this for a few days and then check it. Some people paint the clover at this stage with green food colouring to enhance the colour, but it's up to you.

Now you have your clover, you can choose to:

- laminate it into a bookmark
- mount it in a picture frame (use acid free paper and glue)
- get a clear resin kit and turn it into a paperweight
- mount it in a jewellery piece, such as a resin pendant.

PS: When I got back in touch with Sniper recently to give him a heads up about the chapter that mentions him, goosebumps arose when he told me this about his four-leaf clover: "I still have the clover by the way. Never found one since. I always look."

TIPS FOR A BEE-FRIENDLY GARDEN

Bees, like humans, need a wide variety of foods in their diet to be healthy. By providing a good mix of pollen and nectar producing plants for them (and gardening organically), you will create an environment that they – and you – can thrive in.

When planning your garden, keep in mind that bees need something to eat all year round, so select plants for your climate that will flower through different seasons. They'll also need something to wash it all down with. You can provide a safe place for them to sip water by providing wet sand or a little water saucer filled with marbles or rocks, this allows the bees to sip without the risk of falling in and drowning.

You don't need to plant everything in the list below, just a couple will offer bees a lovely restaurant option in your garden. Take into account the plant species that will thrive best in your climate, soil and conditions – and avoid planting anything that is an invasive species in your area.

Here are some top picks the bees will be grateful for.

ANNUALS

Alyssum (Sweet Alice) – *Alyssum maritimum*
Basil (Summer) – *Occimomum Basilicum*
Basil (Holy/Tulsi) – *Ocimum sanctum/Ocimum tenuiflorum*
Bok Choy (Pak Choy) – *Brassica rapa. sp.Chinensis*
Borage – *Borago officinalis*
Calendula – *Calendula officinalis*
Cockscombs – *Celosia sp.*
Clover – *Trifolium sp*
Coriander/Cilantro – *C. sativum*
Cosmos – *Cosmos bipinnatus*
Fan flower – *Scaevola aemula*
Leeks – *Allium ampeloprasum L.*
Mizuna – *Brassica Rapa subsp.nipposinica*
Nasturtiums *Tropaeolum*
Onions – *Allium cepa*
Pumpkin – *Curcurbita sp.*
Phacelia – *Phacelia tanacetifolia*
Queen Anne's Lace – *Ammi visnaga*
Salvia – many species including *Salvia uliginosa*
Sunflowers – *Helianthus*

PERENNIALS: FLOWERS, SHRUBS AND TREES

Abelia – *Abelia grandiflora*
Banksia – multiple species
Basil (perennial) – *Ocimum americanum*
Bee Bee Tree – *Tetradium daniellii*
Blueberry (multiple species) – *Cyannococcus*
Brushbox – *Lophostemon confertus*
Butterfly Bush – *Buddleja*
Bottlebrush – *Callistemon*

Christmas Bush – *Ceratopetalum gummiferum*
Coastal Rosemary – *Westringia fruticosa*
Dandelion – *Taraxacum*
Echinacea (Purple Coneflower) – *Echinacea purpurea*
Echium – *Echium vulgare*
Eucalypt – Ironbark, Yellowbox, Brushbox
Geranium/Cranesbills – *Geranium*
Grevilleas – Many types including *Grevillea Moonlight*
Hakea/Pincushion tree – *Proteaceae*
Lavender – *Lavendula dentate*
Lemon tree – *Citrus x limon*
Lemon-scented Tea Tree – *Leptospermum polygalifolium*
Lemon balm – *Melissa officinalis*
Lemon myrtle – *Backhouse citriodora*
Lime – *Tilia*
Macadamia – *Macadamia integrifolia, Macadamia tetraphylla*
Melaleuca's including Manuka and Paperbarks
Marjoram – *Oreganum majorana*
Mints – *Mentha x piperita, Mentha x spicata*
Oregano – *Origanum vulgare*
Persimmon – *Diospyros kaki*
Plum – *Prunus salicina*
Portugal Laurel – *Prunus lusitanica*
Raspberry – *Rubus sp.*
Sage – *Salvia officinalis*
Strawberry – *Fragaria sp*
Thyme – *Thymus spp*
Wattle – *Acacia pycnantha*
Winter Savory – *Satureja*
Zinnias – *Zinnia sp.*

In addition to the above plants for bees, some of my other favourite plants to grow are listed below. I feel so lucky to have literally grown alongside these plants, so if you are in a suitable climate (sub-tropical to temperate, or with good microclimates) check these ones out to see if they might suit your garden.

- Rainbow Swiss Chard and Silverbeet – *Beta vulgaris subsp. cicla var. Flavescens.* Such a hardy plant and a great green in pies, on pizza, in a crepe with fetta and tasty cheese and cherry tomatoes, or cooked with ginger, garlic and soy sauce.
- Chinese raisin – *Hovenia dulcis.* A slender tree with the most delicious gnarled fruit that is also good for liver function.
- King White Mulberry – *Shahtoot.* Great for shade, the leaves provide good animal fodder and the berry is long and sweet and white, which means the washing doesn't get stained when birds chew and drop them overhead.
- Carob – *Clifford.* You can read why in Chapter 7.
- Rosella – *Hibiscus sabdariffa.* The young leaves are tangy and the red calyx can be used for tea, jam and cordial.
- Clumping bamboo – *Bambusa oldhamii.* A non-invasive species for windbreaks that also provides edible shoots for humans, leaves for livestock, poles and dyes.
- Lemon myrtle – *Backhousia citriodora.* A beautifully scented leaf that is great for tea, culinary flavor and essential oils.
- Chinese Jujube – Ziziphus jujuba Mill. Slow growing but with a date-like fruit that is delicious dried and also has medicinal benefits.
- Australian bottle tree – *Brachychiton rupestris.* If any tree was created to be hugged, this is the one. It has a fat, bulbous tummy trunk that you just want to wrap your arms around.
- Turmeric – *Curcuma longa.* A lovely fleshy leaf and a rhizome underneath packed with medicinal, culinary and dyeing properties.

RECIPE FOR A BEE MOTEL/NATIVE BEE HABITAT

Like humans, bees have all sorts of preferences for shelter and habitat. Many native bees are ground dwelling, living in tunnels beneath the ground. Other bees live in the thin cavities of reeds and hollow sticks, whilst others prefer drilling into empty stumps or thick bark. By providing habitat in your garden for native bees, you not only help make native insects more resilient, you'll also increase pollination success for you and your neighbours!

Native bees exist across the world and many species have similar housing requirements, here is some basic information to get you started on your journey into discovering the native bees on your continent. The following is about Australian bees, but it's easy for you to discover and house bees native to your region.

AUSTRALIAN BEES
Stingless bee (*Tetragonula & Austroplebeia*): *social, suggested habitat* ■ □
Reed bee (*Exoneura & Braunsapis*): *solitary, suggested habitat* ● <>
Resin bee (*Megachile*): solitary, suggested habitat ● <>
Teddy Bear bee (*Amegilla*): *solitary, suggested habitat* ◊✹
Blue banded bee (*Amegilla*): *solitary, suggested habitat* ◊

Leafcutter bee (*Megachile*): *solitary, suggested habitat* ●
Carpenter bee (*Xylocopa*): *solitary, suggested habitat* ●
Homolictus bee (*Homolictus*): *solitary, suggested habitat* ●

HABITAT CODE

<> Collect 10-20 dry reeds, hollow bamboo cuttings or twigs with pithy interiors (e.g.: lantana, elderberry, reeds, mango). Bundle together and secure with twine. Place the bundle undercover, or stack into round polypipe or a weatherproof timber frame. Fix it so it doesn't swing in the breeze.

■ There are numerous manmade hive styles that will enable you to keep hives of these bees for pollination and/or small amounts of honey. Research the best materials and construction for your area.

□ Retain hollow logs, cap one end if needed and keep in a covered area.

● Drill holes of varying size and length into local native hardwood and softwood timbers and logs. Position in a sheltered area. Don't drill right through when creating the holes, good sizes to try include 6mm x 100mm and 8mm x 150mm.

◊ Form mud brick blocks of local dirt, clays and sand and place in an undercover area (3 parts sand to 1 part clay is a good mix for the blocks, but experiment over the season to see what combo the bees most prefer).

❀ Make a firm sand mound (or pit if in a well-drained area). Mix sand and loam together. Build a shelter over the top.

NATURAL BURN BALM RECIPES

Serious burns should be treated immediately by a doctor, but for
everyday household burns, hold burn under running water for at least
two minutes (15 minutes is better), then dab with apple cider vinegar
and reach for some Aloe vera. *Aloe barbadensis* is a must have plant in
your first aid arsenal. An easy to grow succulent which will thrive in
your rockery, garden or a pot, just cut a thick, fleshy leaf and expel the
cooling plant gel directly onto the burn every few hours. Or have one
of the following balms ready for action.

HONEY BURN BALM
Ingredients:

- 2 tablespoons raw honey (Manuka, Jellybush or another
  highly active medical honey if possible)
- 1 tablespoon coconut oil
- Aloe vera gel

Method:

- Mix together well.

- Apply and then bandage (so everything doesn't get sticky).
- Change dressing twice a day.
- Shelf life approximately two months.

HONEY BURN SALVE
Ingredients:

- 2 tablespoons raw honey (Manuka or Jellybush honey if possible)
- 2 tablespoons coconut oil
- 1 teaspoon beeswax

Method:

- Bring a small amount of water (about 5cm/2 inches) to a gentle boil in a saucepan.
- Add beeswax to a Pyrex cup and place inside saucepan (double boiler method).
- When beeswax has melted, mix in the coconut oil.
- Add honey and whisk till uniform.
- Remove from heat and pour into containers to set. Apply as needed.
- Shelf life approximately one year.

ITCHY BITE BALM RECIPE

Not only does this balm soothe, it also has the most gorgeous scent that some people like to wear as a perfume!

Ingredients: base

- 70 g / 2.5 oz Cocoa Butter
- 60 g / 2.0 oz Shea Butter
- 40 g / 1.5 oz Jojoba oil
- 25 g / 1.0 oz Beeswax

Ingredients: essential oils

- 100 drops each of lemon myrtle, lavender, lemon, tea tree (100 drops = approximately 5mls or 1/6th of an ounce)
- 25 drops of chamomile
- Optional: 10 drops of calendula, 10 drops of Vitamin E

Utensils needed:

- Saucepan.

- 500 ml Pyrex measuring cup.
- A paddlepop stick or something for stirring.
- Small glass containers, preferably dark to protect the balm from light.

Method:

- Add some water to the bottom of the saucepan, so that when you place the Pyrex measuring cup inside (double boiling method) the water comes up to no more than halfway (this is so you don't get water splashes into your balm).
- Place the saucepan on the stove on medium heat as you want it at a slow boil.
- Add the cocoa butter, shea butter, jojoba oil and beeswax to the Pyrex measuring jug, then put the measuring cup into the saucepan with the handle over the edge of the pot so you can access it.
- Give it a stir every now and then.
- While this is melting, measure out your essential oils and keep separate.
- Once the base has all melted, remove from heat and wait a few minutes till it slightly cools (but not too long or it will start going solid). You do this as you don't want the essential oils to evaporate off in the heat. Once the balm base has cooled a little, add in the essential oils and stir well.
- Pour into small, dark containers and allow to cool and solidify, then cap the jars.

Voila! You have just made yourself the most gorgeous natural and effective itchy bite balm...and it smells wonderful too!

ON YOUR FACE JUST USE HONEY, HONEY

Whenever I felt totally shattered after long days and physical hauls and had to pick myself up for another early market, I'd take a shower the night before and make a little concoction for my face. If I had some crystallised honey on hand, I'd just wash my face, then apply, and use it as a natural exfoliator. If I had a tiny bit more time and all the ingredients, I'd make a honey and goat milk face mask.

Ingredients:

- 1 tablespoon honey
- 1-2 tablespoons rice flour
- 1 tablespoon raw goat milk (or whatever milk you have in your fridge)

Method:

- Mix together in a stainless steel bowl, play with the amounts to get a texture that will be gooey and moist enough to stick to your face, but not runny.
- Wash your face, then apply.

- Leave on for 5 minutes (try not to answer the door during this time as you will be quite a sight).
- Remove gently with lots of warm water.
- Dry your face and apply a small amount of Farm Balm.

Feel how soft and smooth your skin is! You can also give your hands a mask at the same time, or any part of your skin that you'd like to renew.

FARM BALM RECIPE

I don't think I ever made a better product than Farm Balm (thought the itchy, lip and sleep balms come pretty close!) and I'd like to share the recipe with you. Why? Because it's great for your skin and good for the planet too. It can be used on heels to hands, face and décolletage, basically wherever you have skin you want to give a bit of love to! It's all natural, all edible and the beeswax acts as a humectant drawing moisture to the skin. Whereas many commercial moisturisers are creams based on water, Farm Balm's base of beeswax and oil means you only need to apply a tiny amount at a time as it won't just evaporate off. It will take a few minutes to soak in but will continue to nourish your skin throughout the day. The following quantity will give you a supply to last you and your family/friends for months.

Ingredients:

- 2 cups (500 ml / 1 pint) organic Coconut oil
- 1 cup (250 ml / 8.5 oz) organic Jojoba oil
- 1 cup (250 ml / 8.5 oz) organic Olive oil
- 8 tablespoons (47 g / 1.6 oz) Beeswax

- 60 ml / 2 oz organic Rosehip oil
- 60 ml / 2 oz organic Evening Primrose oil
- 40 drops chamomile in jojoba
- Optional: measure the final mix and add 1% by volume of Vitamin E. eg: 1 ml Vitamin E to 100 ml of recipe

Method:

- Find a saucepan large enough to fit an 8-cup capacity Pyrex dish and cover the bottom with water (when you place the Pyrex in it, you don't want the water past the half way market or it might spill into your balm ingredients. This will be your double boiler. Set it on medium temp.
- Add the coconut, jojoba, olive oils and beeswax to the Pyrex, stir every now and then with a clean wooden stick or something you don't mind getting wax on.
- Once melted, add the Rosehip and Evening Primrose and stir in.
- Take off the heat and add the Chamomile and Vit E. Stir and pour into sterilised pots and jars.

Note: Farm Balm will be solid in cool weather and runny in hot. A little goes a looooong way. As there is no additional preservative, use within 12 months of making and try and use a spatula so you're not adding germs in via your fingers.

SPRAIN AND BRUISE BALM RECIPES

Comfrey (*Symphytum officinale*), is a fleshy leaved perennial herb with multiple uses in a garden and home. It is a great activator of compost, the bees love the small bell-shaped flowers, it's an important forage for chickens, can act as a border for your garden to stop weeds entering, can be used as a fertilising mulch or steeped in water for five weeks and applied as a fertilising tea for other plants.

In the old days, Comfrey was known as "knitbone" for its ability to speed the healing of broken bones, sprains and bruises. It can be used on livestock and humans alike. Modern science has shown one of its most active components is allantoin which stimulates cell growth and is an anti-inflammatory.

Warning: Plant it in the best spot the first time, as it's super hard to remove once it gets going. Wear gloves to avoid the fine, irritating hairs.

COMFREY POULTICE
Ingredients:

- Fresh cut comfrey leaves (approx 4-8 large leaves) and water

- A handful of either cornflour/bran/flour or some other binding agent to make it pasty

Method:

- Roughly chop the comfrey leaves.
- Add to blender with 1/4 to ½ cup of water.
- Add handful of flour to make a paste (you don't want it drippy).
- Add the mixture to gauze pads or a cloth and apply to skin. Wrap with cling wrap if you are worried about leakage. Change and make a fresh poultice after about 4 hours.

EASY COMFREY BALM
Ingredients:

- 30 g / 1 oz comfrey leaves (2-4 leaves)
- 300 ml / 10 oz olive oil (or sunflower)
- 30 g / 1 oz beeswax

Method:

- Finely chop the comfrey leaves (stems and all).
- Add oil to pan and set to medium heat.
- Add comfrey, simmer gently, stir constantly for 40 minutes.
- Pour the mixture into a cheesecloth bag. Squeeze and strain the liquid back into the pan. Discard leaves.
- Put the strained oil on low heat and add the beeswax. Stir until the beeswax melts
- Pour the mixture into small jars to cool.
- Once cool, apply as needed. Will keep in the fridge for 3 months.
- Optional: the balm will have a grassy/earthy smell, so add a drop or two of lemon essential oil or oil of your choice once the beeswax has melted.

BEESWAX LIP BALMS

Audrey Hepburn said, "For beautiful lips, speak kind words," and I couldn't agree more. But if you are going to put something on yours, make sure it's edible!

HONEY FIX MY LIPS

Ingredients (enough to make 20 tubes):

- 40 g / 1.5oz jojoba oil
- 22 g /0.8 oz cocoa butter
- 19 g / 0.7 oz beeswax
- 11 g / 0.4 oz shea butter
- 7 g / 0.25oz coconut oil
- 4 gm / 0.15 oz carnauba wax
- Essential oils of your choice

Method:

- Into a two cup Pyrex measuring cup add the jojoba, cocoa butter, beeswax, coconut oil and carnauba wax.
- Into a separate 1 cup Pyrex measuring cup add the shea butter.
- Place each cup in a separate saucepan, with water in the base of the saucepans to act as a double boiler (but not so full the water will contaminate your balm ingredients).
- Gently warm until all ingredients have melted. Once they have, add the shea butter into the main mixture.

Remove from heat, and stir in your choice of essential oils (10 drops per 100 ml/3 oz of balm mixture). I used to use a mix of Geranium bourbon, lemon, sandalwood, tea tree and patchouli

HONEY SPEARMINT LIP BALM

Ingredients:

- 4 tablespoons Jojoba oil
- 4 teaspoons grated beeswax
- 2 teaspoons honey
- 20 drops organic spearmint essential oil

Method:

- Into a two cup Pyrex measuring cup add the jojoba, beeswax and honey.
- Place the cup in a saucepan, with water in the base of the saucepan to act as a double boiler (but not so full the water will contaminate your balm ingredients).
- Gently warm until all ingredients have melted.
- Remove from heat, stir, stir in the essential oil
- Scoop into lip balm pots and allow to cool. Cap. This is a soft balm.

GOLDEN HEALING PASTE

Turmeric is anti-inflammatory, anti-cancer and an immune booster. It's easy to grow in a large pot or in the ground, and you can freeze it for use in smoothies, curries, casseroles, rice and add it to so many things. Taking it with black pepper markedly increases the active ingredients – curcumin – absorption in your body.

Ingredients:

- ½ cup turmeric powder
- 1 cup water
- 1.5 teaspoons ground black pepper
- 5 tablespoons coconut or olive oil
- Raw honey to taste/moisten

Method:

- Mix the water and turmeric in a small pan and heat gently.
- Stir for up to 10 minutes until it forms a thick paste, add honey if it is too thick.

- Add the oil and black pepper and stir until well mixed.
- Cool, bottle and then refrigerate for up to 2 weeks.
- Use ¼ tsp up to 3 times a day, have straight like a medicine or add into smoothies, soups and curries.

GROWING TURMERIC

Turmeric is a relatively easy plant to grow in tropical, sub-tropical and temperate climates with good rainfall. In colder climates you will need to grow the plant in a greenhouse, or in a large pot (at least 30cm/12 inches deep and wide) kept indoors in a very light, sunny, warm spot. You can also grow it outside during the summer, then pot it into a container so it can holiday inside during winter. In all but the warmest climates the leaves above will die off in winter, then you know it's a great time to harvest! It likes a loamy, well drained soil that is loosened prior to planting. Plant turmeric 25-40cm / 9.8in-15.7in apart. Each year at harvest, take some rhizomes for eating and freezing and leave some in the ground to regrow.

GERMAN POTATO PANCAKES
(REIBEKUCHEN)

I first learned this simple dish in Year 8 German class and it remains my Dad's favourite meal, great with a salad or just as a salty Sunday afternoon treat.

Ingredients:

- 1 kg / 2.2 lbs potatoes peeled and grated
- 1 onion grated or chopped very finely
- 1 egg, lightly beaten
- 10 tablespoons plain flour (or gluten free substitute)
- Vegetable oil
- Salt

Method:

- Place grated potato into a clean tea towel and squeeze out

excess moisture (this will ensure your end product is crispy).
- Put potato into a mixing bowl with onion and mix together.
- Stir in flour to coat and then add egg.
- Heat oil in a frypan on medium high heat.
- Gently place spoonfuls of mixture into pan and flatten mixture gently with spatula into pancake shape.
- Fry both sides until crisp and golden.
- To serve: sprinkle with salt and serve with salad (it also goes well with apple sauce or chilli jam).

MUM'S CROWD-PLEASING PANEER

This is delicious on the day or over the next few days on a burger. You'll need:

- Enough fresh milk (goat or cow) to fill your tallest pot two thirds full
- Up to 6 cups of yoghurt OR 6 tablespoons lemon juice or white vinegar

Method:

- Bring the milk to the boil, stirring well so it doesn't burn on the bottom (be super careful it doesn't bubble out of the pan).
- As soon as it boils, remove from heat and stir in three quarters of the yoghurt or lemon juice/vinegar. Stir slowly and gently in the same direction until lumps begin to form. If no lumps form after a few minutes, return briefly to heat or add more yoghurt/lemon juice/vinegar.

- You should notice the mixture separating into whey (greenish liquid) and the curds of cheese.
- Spread cheesecloth, a couple of layers thick, into a large colander and gently spoon the curds into the colander to drain. (You can also pour the small bits in).
- Make a bag of the cheesecloth and twist it closed so the curds can't come out, then run it under cold water to rinse and stop the cooking process.
- You can either hang the bag to drain, or place a weight on top to force the liquid out faster (this will give you a firmer cheese that is great to fry or BBQ!).
- Cut the cheese into either cubes (for curries like Palak Paneer/Paneer with Peas and Potato, or to serve with lemon olive oil, salt and pepper. Or, cut it into steaks for frying. If having as steaks, make a marinade to spoon over it.

Here's an idea for the marinade, based on one which I saw chef Kurma Dasa make at Bent on Food in Wingham. He has amazing vegetarian recipe books if you'd like to check them out.

Ingredients:

- 6 tablespoons maple syrup
- 2 tablespoons tomato paste
- 2 tablespoons soy sauce (or good quality Tamari)
- 2 tablespoons Dijon or seed mustard
- 4 tablespoons lemon juice
- 6 tablespoons water
- ½ tsp cayenne pepper (or to taste)

Method:
Place all ingredients in a small pan and stir over a medium heat. Continue stirring until marinade has reduced slightly and is thick and syrupy. Spoon over the paneer and enjoy!

PHILOMENA YODAPOPE'S SAUERKRAUT

I hope I haven't put you off fermented foods – they're actually wonderful for your gut and overall health, and wonderful on the tastebuds! This is PhilomenaYodaPope's heavenly way of sauerkrauting.

Ingredients:

- 2 kg / 4.4 lbs chopped cabbage (fine or coarse – your choice)
- 3 tablespoons sea salt

Equipment:

- Ceramic crock or large glass container (4 litre capacity)
- China or glass plate that fits snuggle inside the container

Method:

- Place the cabbage in layers into a large bowl, sprinkling salt over it as you put it in.
- Pound the cabbage with a sturdy kitchen implement until the juice is released. Go on, keep pounding.
- Pack the cabbage into the crock and cover with a plate that fits snugly inside.
- Place a clean weight on top and press down on this to help force water out of the cabbage.
- Continue doing this every few hours until the brine rises above the plate (this can take up to 24 hours as the salt draws the water out of the cabbage slowly).
- If the brine does not rise above plate level by the next day, add more salt (dissolve 1 tablespoon of salt in 250 ml water).
- When the brine has risen above the cover, cover with a clean cloth and leave at room temperature for 7 to 10 days.
- Taste for your own preference and when happy, bottle and refrigerate to stop fermentation.

THIS IS NOT A CHICKEN BURGER RECIPE

There will be no chicken tonight, but there will be some yum!

Ingredients:

- As many big mushrooms (Portobello's) as you can eat
- Garlic (a small amount, minced)
- Cooking Oil
- Breadcrumbs (or gluten-free alternative)
- Flour (or gluten-free alternative)
- One egg, lightly beaten
- Rolls and hamburger condiments: lettuce, pineapple, tomato, cheese, sauce of your choice

Method:

- Turn the oven on to 180 degrees celcius and line a baking tray.
- Coat each mushroom in flour.
- Top each mushroom with a small amount of garlic (and any

other toppings you like such as camembert, brown rice, cheese, herbs etc).
- Dip the mushroom into the beaten egg and then into the breadcrumbs.
- Place on the baking tray and cook until the mushroom is tender and the crumbs golden.
- Assemble your burger and enjoy!

ROSELLA CORDIAL RECIPE

We're talking about Rosellas the plant (*Hibiscus sabdariffa*) here, not Rosellas the colourful Australian birds…so no jokes about there being feathers in your cordial. Rosella plants are perennial in the tropics but in temperate areas will die off just before winter so you will need to save the seed. You can eat the young green leaves and they'll add a tang to your salad, and then when the red calyx forms get ready to indulge in a Vitamin C rich tea. Or…you can sweeten it and make an amazing jam that goes perfectly with Camembert, add preserved rosellas to champagne, prepare a syrup for desserts or the following Rosella Cordial recipe that you can dilute with plain or sparkling water, or add to alcoholic cocktails for something really special.

Ingredients:

- Rosellas (wild hibiscus)
- Sugar (1 kg / 2.2 lbs per litre of the strained juice)
- Lemons (2 lemons per 1 litre / 2 pints of cordial), juiced and strained
- Citric acid (1 tablespoon per 1.5 litres/1.5 quarts of cordial)

Method:

- Wash rosellas and fill your largest saucepan with them up to about the two-third mark, cover with water.
- Bring to boil and simmer gently until the red has mostly faded from the calyx into the water.
- Strain the mixture through a sieve (you can either compost the used rosellas, or take them off the seeds and use in a berry pie).
- Measure the strained liquid so you know how much you have.
- Clean the saucepan and add the liquid back in…and add 1 cup of sugar for every litre of the liquid.
- Heat gently until the sugar is dissolved, stir often.
- Once the sugar is dissolved, bring the mixture to the boil for 2 minutes, then take the pan off the heat.
- Add the strained juice of the lemons (approx. 1 to 2 lemons per litre of the liquid
- Stir in the citric acid (1 tablespoon per 1.5 litres of cordial).
- Bring briefly back to the boil.
- Pour into clean, dry, sterilised bottles and seal while hot.
- Once opened, store in fridge, but otherwise, your cordial should be fine in your pantry for 12 months.

MULBERRY CRUMBLE RECIPE

One of our very first farmstay guests was a lady called Mary Milne. She came when the old mulberry was laden with berries and we had no idea what to do with them. This is her recipe with just a few tweaks.

Ingredients:

Filling

- Good–sized bowl of mulberries (3-6 cups) …. if you don't have mulberries, you can use mixed berries
- 10 apples (peeled, cored, cut into wedges)
- ½ cup white sugar
- 1 teaspoon cinnamon
- 1 cup water

For the crumble (we like a thick crust of crumble, but you can reduce by a third if you prefer a thinner layer):

- 2 cups oats
- 1 cup flour
- ¾ cup brown sugar
- 100 g / 3.5 oz butter, cold and cut into small cubes
- 1 tablespoon cinnamon

Method:

- Put the apples in a saucepan with the water, white sugar and cinnamon. Cook over a medium heat until the apples are soft.
- Drain the apples.
- Put the apples back in the saucepan with the mulberries and cook for another minute.
- Put the mixture into an oven proof dish.

To make crumble:

- Put all the crumble ingredients in a mixing bowl and rub in the butter.
- Sprinkle crumble on top of the mixture.
- Cook at 190 degrees celcius for 20 minutes or until crumble is brown on top.

Serve with ice cream, custard and a smile.

VEGETABLE FRITTERS WITH CHILLI JAM

Is there anything better than seeing a gigantic zucchini you've missed in the garden and knowing there's something you can do with it?

Ingredients:

- 250 g / half a pound of carrot
- 250 g / half a pound of zucchini
- 500 g / 1.1 lbs potato
- Optional: 1 onion (or leek or chopped chives)
- 5 tablespoons chickpea or gluten-free flour
- 2 large eggs, whisked
- 3 teaspoons ground or 4 teaspoons of grated fresh turmeric
- 1 teaspoon black pepper
- Salt and other herbs to taste
- Olive oil (for frying)

Method:

- Peel and grate the carrot, zucchini, potato and onion. Place in a colander lined with a tea towel and squeeze out as much moisture as you can (if you add a little salt to the bowl with the zucchini it will help draw out moisture – and save the potato to grate last so it doesn't turn brown).

- Place grated vegetables in a bowl and add the eggs, turmeric, salt, pepper and herbs if using.
- Heat oil in frypan over medium heat.
- Drop large spoonsful into the frypan, gently flatten and fry both sides until golden brown.

CHILLI JAM

Ingredients:

- 7 medium-sized red capsicums (seeds removed)
- 8 birds-eye chillies (more if you want it hotter)
- A piece of ginger (thumb-sized)
- 10 cloves garlic
- Can of diced tomatoes (400 g / 14 oz)
- 700g / 1.5 lbs white sugar
- 250ml / 8 oz red wine vinegar

Method:

- Finely chop the capsicums, chillies, ginger and garlic (use a food processor to save time).
- Pour into a large pot with the tomatoes, sugar and vinegar.
- Bring to the boil, skimming the surface as needed.
- Simmer gently for 1 hour, stirring occasionally. The jam will need more stirring as it reduces to prevent it sticking or burning on the bottom of the pan.
- After an hour or so the jam should have a gooey but not runny texture. Remove from heat and allow to cool slightly.
- Pour it into sterilised jars and seal.

SIMPLE THINGS TO DO WITH SAGE

Sage is a hardy perennial and can easily be grown in a pot or in the ground...just don't give it too much water! It likes full sun and regular pruning to keep it from getting spindly.

There are so many things you can do with sage.

- Fry the leaves as a snack.
- Add fresh leaves to omelettes.
- Chop finely and add to butter to make a sauce for gnocchi or ravioli. You'll need 150gm unsalted butter and 3 tablespoons fresh sage leaves. Melt a small amount of the butter and gently fry the sage in it. Remove the sage, add remaining butter until it melts, then stir the sage back in. Remove from heat, add a squeeze of lemon juice and serve over pasta of your choice.
- Infuse a jar of honey by adding dried sage (poke it down to the bottom of the jar using chopsticks or a knife as you need it totally immersed). After two weeks, strain it out. Use the honey to add to marinades, salad dressings and as a tasty topping on toast.

- Make a sage tea when you have a headache or cold. Just steep leaves in boiling water, and add honey to taste.
- Dry, then chop a tablespoon of sage and a tablespoon of rosemary. Add to 2 tablespoons of rock salt and 2 tablespoons of black peppercorns. Grind with a mortar and pestle (or blitz in a blender). Store in glass. Sprinkle the salt over potatoes; mix with garlic and butter for a delicious garlic bread; or bottle it up and give as gifts.

BEETROOT RELISH RECIPE

Enjoy this relish in sandwiches, on burgers and as a great addition to cheese and crackers!

Ingredients:

- 1 kg /2.2 lbs beetroot
- 2 brown onions
- 2 granny smith apples
- 4 tablespoons orange juice
- 1 cup brown sugar
- 1 cup red wine vinegar
- ½ teaspoon sumac
- ½ teaspoon ground cloves
- ½ teaspoon salt

Method:

- Trim, wash and cook beetroot in a large pot of boiling water for 20 minutes. Drain, let cool slightly and dice

roughly into bite-sized or tiny cubes (wear gloves).
Set aside.

- Combine remaining ingredients (except beetroot) in a large
  saucepan and bring to the boil, stirring often. Reduce heat
  and simmer for 10 minutes, or until apple is tender.
- Add beetroot back in and simmer for 40-50 minutes. Ladle
  into sterilised, hot jars (just out of the dishwasher is good).

SIMPLE HONEY HALOUMI FRY

A delicious light lunch to serve with salad. Super quick so it gets the big thumbs up!

Ingredients:

- Block of haloumi cheese cut into finger-thick steaks
- 2 tablespoons pine nuts
- 1 tablespoon of honey
- Handful of fresh basil
- Cooking oil

Method:

- Add a small amount of oil to a large fry pan and heat on medium.
- Add haloumi "steaks" to hot pan in a single layer.
- Cook until golden brown on one side. Then turn haloumi over and add pinenuts to the pan to toast well.

- When both are golden brown, drizzle with honey and add basil to pan.
- Stir briefly and then remove onto a serving plate.
- Serve and enjoy.

FIVE STAR FUSS FREE CHUTNEY

This is my friend Michelle's signature chutney and loved by all... it has the right balance of sweet and savoury and goes perfectly with any hot or cold meal... cheese on toast, cold meats sambos, roast dinner or on the side of your curry. One of the best things about this chutney – besides the flavour – is that you only have to peel the onions and core the apples as the tomatoes and apples go in with their skins on. You will love it!

Note from Michelle, CFO (Chief Flavour Office) at For Flavours Sake:

This is a really large batch that I make for market and usually yields about 2 litres of chutney, but simple halve the recipe to make at home. Have fun with it... add flavours you like to eat. I like Indian flavours so I will add garam masala, coriander & cumin. Coriander & Fennel seeds also add a lovely aniseed flavour too. You now have the 'foundation' of a good chutney... now have a play with it. Add the spices you love & make it your own.

PS: Keep an eye on it... especially towards the end as it can start to stick to the bottom of the pan. Also, taste along the way and add bits and pieces if you think the flavours need adjusting. Enjoy!

Ingredients:

- 2 kg tomatoes
- 1.5 kg green apples
- 1 kg brown onions
- 250 g sultanas
- 1 kg raw sugar (you can lessen the raw & add some brown for a deeper flavour)
- 1 litre / 2.11 pints vinegar (you can use 1/2 brown & 1/2 apple cider, but whatever you have will work)
- 6 cloves of garlic
- 1 inch piece of grated ginger
- 1/2 teaspoon cayenne pepper
- 1 tablespoon white pepper
- 2 tablespoon curry powder
- 2 tablespoon salt
- 1 tablespoon cloves
- 4 dried chillies (optional)

Method:

- Dice tomatoes, apples and onions and place in a large heavy saucepan.
- Add the balance of ingredients and combine over high heat until the mix boils.
- Reduce the heat and simmer for approximately 5 hours, or until the mix has reduced and thickened.
- Ladle into sterilized jars and cool prior to storing.
- This delicious chutney will only improve with age and will keep for up to 12 months.

SPINACH AND RICOTTA DUMPLINGS

With a garden full of silverbeet and a sister full of culinary genius, this recipe makes for a delightful dinner!

Ingredients:

For the dumplings:

- 250 g / ½ lb ricotta
- 2 eggs
- ½ red onion
- 80 g / 3 oz grated cheddar cheese
- 30 g / 1 oz grated parmesan cheese
- 80 g / 3 oz washed spinach (or kale/silverbeet/chard leaves)
- 80 g / 3 oz plain flour (regular or gluten free)
- ½ teaspoon baking powder

For the sauce:

- 700 ml / 25 oz jar of passata
- ½ cup sundried tomato slices

- ½ red onion finely chopped
- 4 tablespoons tomato paste
- 3 cloves garlic finely chopped
- 1 teaspoon chilli flakes (optional)

To serve:
¼ cup chopped chives
Parmesan cheese, extra, grated.

Method:

- Preheat the oven to 180 degrees celcius / 350 Fahrenheit.
- For the sauce – put all ingredients into a large oven proof baking dish or pan and mix gently.
- For the dumplings – put all ingredients except the flour and baking powder into a food processor and blend until chopped and well combined.
- Transfer dumpling mix to a mixing bowl and add the flour, baking powder and a pinch of salt and pepper. Mix gently until just combined – but do not over mix and do not compress the mixture as much as possible.
- Use a small ice cream scoop or tablespoon measure to gently scoop balls of the mixture and place straight into the sauce. (Be gentle with the mixture to keep the dumplings light and fluffy).
- Bake uncovered for 30 -35 minutes.
- Sprinkle with chives and extra parmesan to serve.

SWEET STUFF

Okay, let's indulge!

MULBERRY JAM

Ingredients:

- 1.75 kg / 3.8 lbs mulberries or any mix of berries of your choice
- 2 kg / 4.4lbs caster sugar
- 3 tablespoons powdered pectin (Jamsetta) or you can use lemon juice
- 1 tablespoon unsalted butter

Method:

- Place berries in a large saucepan and mash them (potato masher works great) gently while on low heat.
- Add pectin and sugar, stirring until dissolved.
- Add the butter and bring to the boil. Boil for 4 minutes.

- Remove from heat, cool for 1 minute and ladle into sterilised jars. Seal immediately.

Serve on toast...or bake some camembert and pop the jam on top!

JACK'S FRENCH TOAST

Ingredients:

- 160 ml / 5 oz milk (any type)
- 3 eggs
- 85 g / 3 oz sugar
- 1 teaspoon vanilla essence
- Butter (for frying)
- Bread (thick sliced)

Method:

- Whisk eggs and add to milk, sugar and vanilla essence. Stir to combine.
- Heat butter in a frypan on medium heat.
- Dip bread slices into the mixture, lay gently in the frypan.
- Cook until golden on each side, adding more butter as needed to prevent the frying pan from drying out.
- Serve: with maple syrup, brown sugar & lemon juice, or vanilla-infused honey.

VELVETY CAROB/RAW CACAO MOUSSE

Is there a more decadent, healthy dessert than this? Smooth and velvety, I eat it by the guilt-free spoonful.

Ingredients:

- 1 or 2 ripe avocados, flesh only
- 1 tablespoon raw cacao powder or carob powder
- 3 tablespoons honey
- Optional: pinch of salt and/or a drop or two of vanilla

Method:

- Use a hand blender to whip this into a mousse.
- Add more honey and/or cacao for taste/texture and re-blend until smooth.
- Enjoy on its own or with a side of berries or orange pieces.

SECRET RECIPE SLICE

It mightn't be particularly healthy, but gee it'll make your tastebuds happy! This is the recipe even the CWA ladies begged for! But please remember, there's no need to coif the dates (see Chapter 23 to remind you why).

Ingredients:

- 2 cups (500 g / 17 oz) unsalted butter
- 4 cups chopped dates
- 4 cups brown sugar
- 2 teaspoons ground ginger
- 4 eggs, beaten
- 4 cups self-raising flour
- 4 tablespoons honey
- 2 cups chopped walnuts

Method:

- Preheat oven to 180 degrees celcius. Line two high-sided baking trays with non-stick paper.
- Put the butter, dates, brown sugar and ginger to a large saucepan. Cook on medium low heat, stirring often until the dates are very soft and the mixture is brown.
- Stir in honey.
- Remove from heat and allow to cool slightly.
- Stir in half the flour, and then the egg (you don't want the egg turning into scrambled egg in the heat, so add it last). Add remaining flour and stir well.
- Gently fold through the walnuts.
- Use a spatula to help smooth the mixture evenly into the pans
- Depending how big your trays are, cook for 15-30 minutes until golden brown and a skewer comes out clean when tested.
- Remove from oven and place on racks to cool.
- This stores super well in the fridge and freezer

SLEEP BALM

Feeling tired but your mind is racing? Are your own plans, dreams, worries, big ideas and loves making it hard for you to get off to sleep?

This is the balm I make to get myself off to dreamland.

Combined with a relaxing tea of lemon balm, lavender, chamomile, vanilla and rose petals, it's a beautiful way to help you rest and renew overnight.

Apply to wrists and temples at bed time, take deep breaths, and breathe nature and calm in.

Sweet dreams.

Ingredients:

- 40 g / 1.4oz jojoba oil
- 18 g / 0.6 oz beeswax
- 15 g / 0.5 oz cocoa butter
- 10 g / 0.35oz coconut oil
- 5 g / 0.18 oz carnauba wax
- 60 drops of essential oils per 100ml of base.
- I used a mix of organic mandarin, lavender true, marjoram sweet, lime, palmarosa and myrrh.

Method:

- Into a two cup Pyrex measuring cup add the Jojoba, cocoa butter, beeswax, coconut oil and carnauba wax.
- Place the cup in a saucepan, with water in the base of the saucepan to act as a double boiler (but not so full the water will contaminate your balm ingredients).
- Gently warm and stir until all ingredients have melted.
- Remove from heat, and stir in the essential oils.
- Pour into small jars or wind up lip balm or sport tubes.
- Allow to cool. Cap. Use within 12 months to ensure the essential oils are at their best.

Zzzz…

ACKNOWLEDGMENTS

To Nicolette and all the amazing, caring backpackers who brought life and laughter to the farm, thank you. To all the great guests who supported us by coming to stay, thank you. To all the market customers who continue to choose to support small farms rather than faceless global corporations, thank you. To all the bus tour groups who laughed at my jokes and encouraged us with kindness, thank you!

To Gloria Kempton and Kathy Meyer for your intellect, expertise and encouraging emails, I wouldn't have dared finish this book without your support. And to Sophia Barnes, Samantha Miles, Roz Hopkins and Geoff Whyte, thank you for your words along the way.

To Helen and Jack, thanks for your eagle eyes on the draft manuscript and to Jill McIttrick for making it, and many markets, better. Without Fran McCowan, the recipes would never have been wrestled into shape and without Frank and Pat I never would have learned to laugh so much, through so much and with so much.

To Gayna Murphy, you do art with all heart, and Susan Lowick, thanks for making me laugh so I didn't realise you were taking the pic.

To Cathy and Barry, Chris and Lisa, Jodie and Stuart, Helen and Bruce, Trish and Evan, Nicole and Sean, Leanne, Susie Juice, Daryl,

Hayes, Krissy, Jenn, Dana, Shona, Christine and Phil, Lisa and Ellie, Al and Mel – thanks so very, very much for supporting us and coming along for the ride. And to you kind reader, for coming along too.

To the Firefly Book Club for the laughs and lightness and for reminding me of the benefits of having a night off with a good book, good wine and good people. And to the great people at the Nabiac Old Bank Centre – for always making my day.

And to my you-are-so-loved-it's-just-that-no-words-will-ever-be-able-to-describe-it children and Andrew, thanks for putting up with the hectic and just for being you.

Finally, to all the animals, but especially Kiara, Mickey, Fun Factory, Coco Pops, Daffy and Nick; Rembrandt and Orange Blossom; Rocky, Moonlight, Magic, Cliffhanger, Bucket Chicken and Chicken a'la Orange; Hollywood, Soda, Cindy and Cody; Jasper, Jarrah, Cienta, Brutus, Bravado and Spring; Billy, Betsy, Ivy, Tracey, Sandy, Bubblegum, Vanilla, Princess, Panda, Henry, Soloette, Gigi, Gwen, Peaches and Cream; Rammie, Yoda, Jedi, Fat Max, Mikey, Likey, Molly, Polly, Dear, Queenie, King, Solo and Black Jack; feral CatCat (aka Shadow, Michael, Cat-of-no-value); and the never to be forgotten bees, bugs, plants, landscape, microbes, creek, seasons, sunrises and sunsets who we shared this life with, thank you for opening my eyes, heart, senses and soul. Big love and best of bestest wishes. Ever. Whenever. Forever.

* * *

If you enjoyed this book (and yes, the names and characteristics of some people and animals were changed to protect privacies – and yes, these are my recollections and interpretations and not those of a surveillance camera), it would be lovely if you would consider leaving a review at GoodReads, Amazon and other book lover sites.

If you're still interested in farming after reading about these agricultural adventures (do hope so because the earth needs great people to connect with the land!), Anna and Andrew have co-written a much more serious and informative book called "Small Farm Success

Australia – How to Make a Life and a Living on the Land"....of course they interviewed 25 farmers who have made a real success of it!

They have also interviewed farmers from the US, Canada, UK and Europe for a series of international small farming eBooks more relevant to those countries.

Visit   www.smallfarmingsuccess.com   or   tootle   over   to www.honeyfarmdreaming.com to find out more.

Bee well people. xox

CPSIA information can be obtained
at www.ICGtesting.com
Printed in the USA
LVHW040145050522
717841LV00052B/2516